D0712357

SUPPLY CHAIN ANALYSIS

A Handbook on the Interaction
of Information, System and
Optimization

Recent titles in the **INTERNATIONAL SERIES IN**
OPERATIONS RESEARCH & MANAGEMENT SCIENCE
Frederick S. Hillier, Series Editor, *Stanford University*

Sethi, Yan & Zhang/ *INVENTORY AND SUPPLY CHAIN MANAGEMENT WITH FORECAST UPDATES*

Cox/ *QUANTITATIVE HEALTH RISK ANALYSIS METHODS: Modeling the Human Health Impacts of Antibiotics Used in Food Animals*

Ching & Ng/ *MARKOV CHAINS: Models, Algorithms and Applications*

Li & Sun/ *NONLINEAR INTEGER PROGRAMMING*

Kaliszewski/ *SOFT COMPUTING FOR COMPLEX MULTIPLE CRITERIA DECISION MAKING*

Bouyssou et al/ *EVALUATION AND DECISION MODELS WITH MULTIPLE CRITERIA: Stepping stones for the analyst*

Blecker & Friedrich/ *MASS CUSTOMIZATION: Challenges and Solutions*

Appa, Pitsoulis & Williams/ *HANDBOOK ON MODELLING FOR DISCRETE OPTIMIZATION*

Herrmann/ *HANDBOOK OF PRODUCTION SCHEDULING*

Axsäter/ *INVENTORY CONTROL, 2nd Ed.*

Hall/ *PATIENT FLOW: Reducing Delay in Healthcare Delivery*

Józefowska & Węglarz/ *PERSPECTIVES IN MODERN PROJECT SCHEDULING*

Tian & Zhang/ *VACATION QUEUEING MODELS: Theory and Applications*

Yan, Yin & Zhang/ *STOCHASTIC PROCESSES, OPTIMIZATION, AND CONTROL THEORY APPLICATIONS IN FINANCIAL ENGINEERING, QUEUEING NETWORKS, AND MANUFACTURING SYSTEMS*

Saaty & Vargas/ *DECISION MAKING WITH THE ANALYTIC NETWORK PROCESS: Economic, Political, Social & Technological Applications w. Benefits, Opportunities, Costs & Risks*

Yu/ *TECHNOLOGY PORTFOLIO PLANNING AND MANAGEMENT: Practical Concepts and Tools*

Kandiller/ *PRINCIPLES OF MATHEMATICS IN OPERATIONS RESEARCH*

Lee & Lee/ *BUILDING SUPPLY CHAIN EXCELLENCE IN EMERGING ECONOMIES*

Weintraub/ *MANAGEMENT OF NATURAL RESOURCES: A Handbook of Operations Research Models, Algorithms, and Implementations*

Hooker/ *INTEGRATED METHODS FOR OPTIMIZATION*

Dawande et al/ *THROUGHPUT OPTIMIZATION IN ROBOTIC CELLS*

Friesz/ *NETWORK SCIENCE, NONLINEAR SCIENCE and INFRASTRUCTURE SYSTEMS*

Cai, Sha & Wong/ *TIME-VARYING NETWORK OPTIMIZATION*

Mamon & Elliott/ *HIDDEN MARKOV MODELS IN FINANCE*

del Castillo/ *PROCESS OPTIMIZATION: A Statistical Approach*

Józefowska/*JUST-IN-TIME SCHEDULING: Models & Algorithms for Computer & Manufacturing Systems*

Yu, Wang & Lai/ *FOREIGN-EXCHANGE-RATE FORECASTING WITH ARTIFICIAL NEURAL NETWORKS*

Beyer et al/ *MARKOVIAN DEMAND INVENTORY MODELS*

Shi & Olafsson/ *NESTED PARTITIONS OPTIMIZATION: Methodology And Applications*

Samaniego/ *SYSTEM SIGNATURES AND THEIR APPLICATIONS IN ENGINEERING RELIABILITY*

Kleijnen/ *DESIGN AND ANALYSIS OF SIMULATION EXPERIMENTS*

Førsund/ *HYDROPOWER ECONOMICS*

Kogan & Tapiero/ *SUPPLY CHAIN GAMES: Operations Management and Risk Valuation*

Vanderbei/ *LINEAR PROGRAMMING: Foundations & Extensions, 3rd Edition*

Chhajed & Lowe/ *BUILDING INTUITION: Insights from Basic Operations Mgmt. Models and Principles*

Luenberger & Ye/ *LINEAR AND NONLINEAR PROGRAMMING, 3rd Edition*

Drew et al/ *COMPUTATIONAL PROBABILITY: Algorithms and Applications in the Mathematical Sciences*

Chinneck/ *FEASIBILITY AND INFEASIBILITY IN OPTIMIZATION: Algorithms and Computation Methods*

* A list of the early publications in the series is at the end of the book *

SUPPLY CHAIN ANALYSIS

A Handbook on the Interaction of Information, System and Optimization

Edited by

Christopher S. Tang
UCLA Anderson School of Management

Chung-Piaw Teo
NUS Business School, National University of Singapore

Kwok-Kee Wei
Faculty of Business, City University of Hong Kong

 Springer

Christopher S. Tang
University of California at Los Angeles
California, USA

Kwok-Kee Wei
City University of Hong Kong
Hong Kong

Chung-Piaw Teo
National University of Singapore
Singapore

Series Editor:
Fred Hillier
Stanford University
Stanford, CA, USA

Library of Congress Control Number: 2007935247

ISBN-13: 978-0-387-75239-6   e-ISBN-13: 978-0-387-75240-2

Printed on acid-free paper.

9 8 7 6 5 4 3 2 1

springer.com

Contents

Preface vii

Supply Chain Configurations of Foreign Cosmetics Companies
Operating in China 1
**Cindy Fang, Frances Gao, David Liu, Christopher S. Tang, Weiwei Wang,
and Tony Wu**

Structural Supply Chain Collaboration Among Grocery Manufacturers 29
Timothy M. Laseter and Elliott N. Weiss

Supply Chain Management in the Chemical Industry: Trends, Issues,
and Research Interests 45
Hong Choon Oh, I.A. Karimi, and R. Srinivasan

Berth Allocation Planning Optimization in Container Terminals 69
Jam Dai, Wuqin Lin, Rajeeva Moorthy, and Chung-Piaw Teo

Merchandise Planning Models for Fashion Retailing 105
Kumar Rajaram

Supply Chain Management in the Presence of Secondary Market 147
Hau L. Lee, Barchi Peleg, Seungjin Whang, and Yan Zou

Global Diffusion of ISO 9000 Certification Through Supply Chains 169
Charles J. Corbett

Risk Management in Global Supply Chain Networks 201
N. Viswanadham and Roshan S. Gaonkar

Trust and Power Influences in Supply Chain Collaboration 223
Weiling Ke and Kwok-Kee Wei

Foreign Direct Investment or Outsourcing: A Tax Integrated Supply Chain
Decision Model   241
N. Viswanadham and Kannan Balaji

Integrating Demand and Supply Chains   261
Puay Guan Goh

Mumbai Tiffin (Dabba) Express   271
Natarajan Balakrishnan and Chung-Piaw Teo

Index   279

Preface

Supply chain academic research has traditionally focused on the interaction of information and processes, with goals of streamlining the various activities within the supply chain to obtain the optimal or desired outcomes. Most of the insights, however, were obtained under the ideal situation where the settings are the most general. Practitioners, on the other hand, have focused on developing cost effective solution strategies on the ground, to adapt to local and industry specific conditions.

The chapters in this handbook are compiled with the intent to bridge this gap.

In the first chapter (Supply Chain Configurations of Foreign Cosmetics Companies Operating in China), Fang et al. examined the issues of supply chain design, from the perspective of foreign cosmetic companies operating in China. They examined the market potential of this industry, and propose three major supply chain configurations in this industry (Direct or Indirect Export of Finished Goods, Outsourced Manufacturing, or Offshore Manufacturing). They further examine the various supply chain issues associated with each configuration. The chapter blends extensive industry surveys with a comprehensive framework and extensive list of issues to spur future research in this area.

In the second chapter (Structural Supply Chain Collaboration Among Grocery Manufacturers), Laseter and Weiss continued this theme, and examined the issue of supply chain coordination among the grocery manufacturers. Most notably, they quantify the opportunity for strategic, multilateral collaboration through a shared distribution network. This moved one step beyond the current focus on CPFR and VMI, which are essentially collaborative efforts focusing on inventory turns and service performance.

Oh, Karimi and Srinivasan (Supply Chain Management in Chemical Industry) looked at the supply chain issues in the chemical industry. The chapter identified several unique characteristics, emerging trends, and operational issues of the industry and proposed new research agenda.

Dai et al. (Berth Allocation Planning Optimization in Container Terminal) examined the issue of supply chain efficiency in the container terminal environment. In this chapter, they focused on the issue of berth planning, a key step in the port operating environment. They propose a class of policies to optimize the use of berthing space, to minimize lateness and delays to vessels, and to achieve the optimal usage of space in the terminal. The model is tested and validated using data from a real container terminal.

In the fifth chapter (Merchandise Planning Models for Fashion Retailing), K. Rajaram developed strategies to improve the accuracies of merchandise testing in fashion retailing. For instance, he found that some of the variations of the sales of a product mix can be explained by store descriptors, and used this observation to develop a new merchandising process. He tested the ideas on a large women's apparel retailer, and a catalog retailer, and reported significant savings using the proposed models.

In the sixth chapter (Supply Chain Management in the Presence of Secondary Market), Lee et al. examined the impacts of online secondary market on the operational issues of a supply chain. This problem has become more prevalent, with the proliferation of E-marketplaces. They characterized the optimal responses under various scenarios (e.g. when manufacturers intervened in the secondary market), and identified the key roles played by the secondary markets in the supply chain.

C. Corbett (Global Diffusion of ISO 9000 Certification Through Supply Chains) looked at the contribution of globalized supply chain to the diffusion of ISO 9000 certification, using data from a global survey. He concluded that the diffusion of ISO 9000 started primarily in Europe, and spread to other countries as the European firms pressured their suppliers to seek ISO 9000 certification. This chapter highlighted the role of supply chain on the diffusion of management practices in the industry.

Viswanadham and Gaonkar (Risk Management in Global Supply Chain Networks) looked at a timely issue – the impact of risk and disruption in supply chain. They developed a framework and classification to handle supply chain risk management issues. They developed mathematical models, built on mapping of exceptions and consequences using fault trees and event trees, to design robust inbound supply chains that are resilient to deviations and disruptions.

W. Ke and K.K. Wei (Trust and Power Influences in Supply Chain Collaboration) looked at the issue of information sharing from a socio-political perspective. They distinguished between two types of trust – competence and benevolence, and five types of non-coercive power, and studied their impact on firm's predisposition to sharing information and know-hows. They identified competence-based trust as more important for know-how sharing.

Viswanadham and Balaji (Foreign Direct Investment or Outsourcing: A Tax Integrated Supply Chain Decision Model) examined the impact of taxes on supply chain planning. More specifically, they analyzed the model by incorporating tax-holidays enjoyed by locating the various stages of the supply chain in FTZs. This has become an important concern for supply chain planning, as FTZ is a strategy being used by many developing countries to attract FDI.

The penultimate chapter in this handbook is a contribution by P.G. Goh (Integrating Demand and Supply Chains). He expounded on the needs to synchronize supply with demand chain, and explained the difficulties in doing so with case studies from firms based in Asia.

In the last chapter, Balakrishnan and Teo (Mumbai Tiffin (Dabba) Express) looked at a marvelous distribution system based in India – a system perfected by a group of illiterate workers, without the aid of modern technology and expert advice. The system relies solely on a primitive coding system, and the reliable railway system in Mumbai.

Supply Chain Configurations of Foreign Cosmetics Companies Operating in China

Cindy Fang, Frances Gao, David Liu, Christopher S. Tang, Weiwei Wang, and Tony Wu
UCLA Anderson School, 110 Westwood Plaza, Los Angeles, CA 90095, USA

Abstract: As foreign cosmetics companies develop strategies for establishing or expanding their presence in China, they need to configure their supply network, distribution channels, and outbound distribution network so as to improve coordination and maximize efficiency. In this paper, we investigate the current supply chain configurations of various foreign cosmetics companies operating in China. In addition, we highlight new research opportunities for supply chain designs.

Key words: Supply Chain Configurations, Distribution Channels, Outbound Distribution, Market Entry Strategies, Alignment, China.

1. Introduction

Ever since the 1978 economic reform and the 2001 WTO status, China has experienced a phenomenal economic growth and a steady increase of disposable income.[1] According to the National Bureau of Statistics of PRC (NBS), out of a population of 1.3 billion people in China, over 500 millions reside in urban cities such as Beijing, Shanghai, and Guangzhou. Also, over 65 million middle class people have an annual household income between US$7,230 and US$60,240, and most of them welcome western culture especially products that offer individualism or status. NBS predicted that the middle class would increase from the current 5 to 45% by 2020. These environmental factors attracted many multinational corporations (MNCs) to establish their presence in China through various means. Some corporations outsourced their manufacturing operations to China so as to exploit the abundant low cost labor force, while others established sales channels in China so as to expand their global sales.

In 2006, China is undeniably one of the most important consumer markets that multinational corporations would like to enter or expand. According to a survey conducted by the American Chamber of Commerce, three out of four U.S. companies reported profitable operations in China and most reported higher profit margins in China than elsewhere in the world (c.f., Forney (2004)). For example, Shanghai GM (http://www.gmchina.com/english/), the flagship joint venture in China, was the monthly sales leader in 2005 among all passenger car manufacturers in China. The sharp increase in sales of Chevrolet and Cadillac in China has helped General Motors to narrow its losses in the first quarter of 2006 to $323 million, compared to $1.3 billion loss in the same quarter a year ago (c.f., Ellis (2006)).

To succeed in a foreign market like China, it is increasingly important for a firm to develop competitive advantages from coordinating various activities along the value chain (c.f., Arnold (2003)). These activities include supply chain activities (procurement, in-bound logistics, manufacturing operations, and out-bound logistics) and marketing activities (product development, marketing and sales, and service). Unfortunately, since there is an urge to rush to market, most companies may not have the luxury to develop a plan to coordinate various marketing and supply chain activities when they enter a foreign market. Consider these failures:

---

[1] National Bureau of Statistics of PRC reported that China's GDP was growing at a rate of 9.5% in 2004 and the per capita annual disposable income of urban households was US$1,200 in 2004.

- *Entering a new market without a good marketing plan can be detrimental.* In 1995, Whirlpool, a world's leader in home appliances, formed a joint venture with a well-established refrigerator manufacturer Snowflake Electric Appliance Co. Ltd in Beijing. While the manufacturing operations were running smoothly, only 39,000 of the total output of 62,000 refrigerators were sold in 1996. The main reason for this dismal performance was Whirlpool's marketing strategy. First, unlike other well-known brands in China such as Haier, LG, Hitachi, Sharp, etc., Whirlpool did not launch a strong advertising or promotion campaign to create brand awareness and brand image. Second, Whirlpool positioned their products at the high-end of the market and sold their products only at major retail chains and hypermarkets such as Carrefour in major cities only. As Whirlpool was a relatively unknown brand, Chinese consumers were reluctant to buy a Whirlpool refrigerator at high price without additional perceived value relative to other well-known brands. Even though Whirlpool adopted the Chinese brand name *Hui Er Pu* that means "cheap and popular," the sales continued to be abysmal. After suffering from major losses, Whirlpool terminated their association with Snowflake in 1997. Currently, Whirlpool is focusing on exporting products (microwave ovens, washing machines, and dryers) made in their remaining facilities in China (c.f., Pan (2005)).

- *Entering a new market without a good supply chain plan can be problematic.* In 1981, due to high manufacturing cost in Taiwan and Korea, Nike decided to source their shoes from various contract manu-facturers in China. Even though Nike was a well-known brand in China and Nike had an excellent marketing strategy, they encountered major supply chain problems in China. These problems include the lack of high quality raw materials (nylon, canvas, rubber and chemical compounds) in China, technology transfer issues, quality issues, production planning and inventory control issues, logistics issues, communication issues, and management issues. Without a clear supply chain strategy, Nike was in a bind. Instead of terminating the association with these contract manu-facturers, Nike was able to enlist ADI Corporation, their former contract manufacturer in Taiwan, to manage their operations in China so as to get their manufacturing operations in China back on track (c.f., Austin and Aguila (1985)).[2]

---

[2] After Nike handled this crisis in the late 80s and weathered the storm about sweatshops and child labor issues, Nike has implemented their supply chain strategy and marketing strategy in China successfully, and Nike is now opening on average 1.5 new specialty stores per day in China (c.f., Forney (2004)).

Recognizing the importance of aligning supply chain and marketing activities, we evaluate the underlying supply chain and marketing activities adopted by different companies when they entered a foreign market like China.[3] To gain a clearer understanding, we focus our study based on major companies in the cosmetic industry for the following reasons:

1. *Market Potential.* The cosmetics market in China has grown from around US$25 million in 1994 to $7.9 billion in 2004, and is expected to expand at an annual rate of about 12% (c.f., Tao (2005) and Fang et al. (2006)). Also, the cosmetics industry has become the fifth largest consumption hotspots on the Mainland China only after real estate, automobile, electronics, and tourism (c.f., Li and Fung Research Report, 2005).

2. *Market Readiness.* In 2004, the per capita annual spending on cosmetics in major cities such as Beijing and Shanghai has reached US$20. As living conditions continue to improve, spending on cosmetics is likely to increase exponentially. Also, other regions such as Tianjin, Zhejiang, Hubei, Jiangsu, Chongqing, etc., are experiencing double digit growth in cosmetic sales (c.f., Li and Fung Research Report (2005)).[4]

3. *Brand Awareness and Consciousness.* According to a market research survey of 34,000 people conducted by IMI in 2005, Chinese consumers are becoming more sophisticated and value-oriented. The top two factors affecting consumers' choice of cosmetics in China are brands and prices, which can be influenced by marketing and supply chain activities.[5] This consumer preference is revealed as major international brands such as Olay by Proctor & Gamble (P&G), Aupres by Shiseido, and L'Oréal Paris by L'Oréal captured the top three spots in the market of skin care products (c.f., Li and Fung Research Report, 2005). Moreover, as local Chinese brands (e.g., Dabao and Yu Mei Jing) focus on the low-end market that accounts for 60% of the cosmetic market, 80% of the profit is captured by the international brands (c.f., Tao (2005)).

---

[3] While Porter (1986) presented various generic international strategies for firms to compete globally, our paper focuses on the issue of coordinating supply chain and marketing activities for a firm to succeed in a new market.

[4] In 2004, Shiseido was pleasantly surprised with the sales of their most expensive skin cream Clé de Peau at US$500 per bottle at 4 high-end department stores in China (c.f., Koehn (2005)).

[5] Other factors were packaging, advertising, buying convenience, others' recommendations, etc.

We conduct our fact findings by searching through the Euromonitor International Databases and publications available in the public domain. Also, we conduct personal interviews in China with seven major international cosmetic companies: Avon, Amway, Estée Lauder, L'Oréal, P&G, Revlon, and Shiseido. We develop a framework for analyzing the information gathered in this research. Based on our analysis of the facts collected from these sources, we show how multinational cosmetic companies configure their supply chains and distribution channels so as to compete with other local and international brands. In addition, we highlight some research opportunities for designing supply chain and distribution strategies that would enable companies to improve their efficiency when entering a new market such as China.

2.   Regional Analysis

The cosmetics market in China can be divided into six major regions: East, Middle, North and Northeast, South and Southwest. Table 1 provides a list of provinces associated with each region and the corresponding characteristics. As shown in Table 1, the coastal cities Beijing (North and Northeast), Shanghai (East) and Guangzhou (South) dominate the market due to strong economic growths in these major cities. These three cities are considered as the first tier cities where the most affluent consumers reside. The second tier cities are primarily the capitals of certain provinces, including Wuhan, Xi'an, Chengdu, Shengyang, Harbin, Nanjing, and Hangzhou. Consumers in the second tier cities have disposable income to purchase luxury goods, but they have less exposure to foreign culture than the consumers in the first tier cities. In other cities, most residents are less educated with lower disposable income and they tend to purchase domestic brands such as Daibao.

The projected growth of the cosmetic market in China is shown in Fig. 1. Observe from Fig. 1 that the projected cosmetic market in terms of projected sales and in terms of compound annual growth rate (CAGR) continues to concentrate in three major regions, namely, East, North and Northeast and South. Specifically, in 2004, the East region accounted for 24% of sales, while the South and the North and Northeast regions accounted for 21 and 24% of sales, respectively. As the three key regions are far apart geographically speaking, it is challenging to develop an efficient supply chain especially when many legal requirements are regional.

Table 1. Overview of different geographical regions in China

Region	Province	Population (millions)	Region highlight
East	Shandong, Jiangsu, Zhejiang, and Anhui	220	Shanghai is the economic center of China and it has the largest population
Middle	Henan, Hubei, Hunan, and Jiangxi	323	Agricultural region
North and Northeast	Heilongjiang, Jilin, Liaoning, Hebei, and Shanxi	226	Beijing is the political center of China
Northwest	Qinghai, Gansu, and Shaanxi	113.5	Undeveloped region
South	Fujian, Guangdong, and Hainan	106	Guangzhou is the second strongest economic center of China
Southwest	Sichuan, Yunnan, and Guizhou	160	Chongqing has a large population. Economic development is behind coastal region.

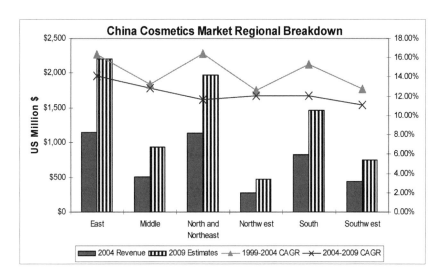

Figure 1. Projected cosmetics market growth in different geographical regions in China

3.   Product Characteristics

Cosmetic products can be classified into nine major categories according to their functions, including: skin care, bath, hair care, color cosmetics, oral care, sun care, baby care, men's grooming, and fragrances. According to the 2005 Euromonitor's database on cosmetics and toiletries in China, skin care category represents 37% market share of the cosmetic market in China, which amounts to over US$2B in sales. The market share of each category is provided in Fig. 2. Within the skin care category, facial moisturizers take up more than 75% of the total sales; facial cleaning creams, whitening creams and anti-acne creams make up the rest of the sales.[6] Other categories of skin care products capture relatively small market share. The key reason is that skin care products are accepted culturally especially whitening products and moisturizers, while color cosmetics products and fragrances are still in the early adoption stage in China.[7] Skin care products also enjoy a higher profit margin than color cosmetics products because of simpler technology and lower consumer taxes.[8] As such, skin care products are much more popular because of their intrinsic appeal: higher value at lower cost.

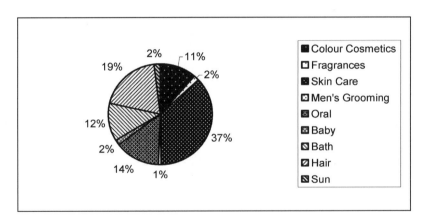

Figure 2. Market share of different cosmetics product categories in China

---

[6] Anti-wrinkle cream did not perform well due to the negative connotation, and hence, many brands are re-naming this subcategory as anti-aging cream recently.

[7] Among all color cosmetics products, lipsticks are by far the most popular item among the Chinese consumers (c.f., Tao (2005). Chinese consumers are still not accustomed to strong fragrances and still prefer subtler scents based on natural ingredients such as herbal essence or fruit extract. For more details about market share of each product category in different regions of China, the reader is referred to Fang et al. (2006) for details.

[8] As of 2004, the consumption taxes associated with skin care and make-up products are 8% and 30%, respectively.

According to a report issued by the Chinese National Commercial Information Center (CNCIC), the most popular brands of skin care products are provided in Table 2. Notice that the skin care products designed specifically for Asian skin such as Aupres by Shiseido are well received in China (c.f., Jones et al. (2005)).

Table 2. Market share of skin care products in China

Brand	Company	2004
Olay	P&G (China)	32.5%
Aupres	Shiseido	14.1%
L'Oréal Paris	L'Oréal	7.5%
Dabao	Beijing Dabao	5.3%
Longliqi	Longliqi Group	3.8%
Yue-Sai	L'Oréal	2.8%
Mininurse	L'Oréal	2.4%
Avon	Avon (China)	0.9%
Others		30.7%

To our knowledge, there is no industry-wide segmentation of different brands of cosmetic products in China. However, we develop our segmentation based on the information gathered from our personal interviews conducted at various multinational cosmetic firms in China. Essentially, most companies use retail price as a way to classify their brands into three major categories: Premium, Mass-premium, and Mass. Premium brands represent products that are sold at retail price above US$40 Mass-premium brands correspond to products with retail price between US$20 and US$40; and Mass brands are referred to products with retail price below US$20. Examples of these three segments of brands are provided in Table 3.

Table 3. Brand segmentation in China

Brand classification	Brand (company)
Premium: retail price is above US$40	Shiseido (Shiseido), SK-II (P&G), Crème De La Mer (Estée Lauder)
Mass-premium: retail price is between US$20 and US$40	MaxFactor (P&G), Revlon (Revlon), Aupres (Shiseido)
Mass: retail price is below US$20	Olay (P&G), Maybelline (L'Oréal)

Besides different retail prices, different classes of brands have different characteristics. For example, Premium brands are usually positioned as "symbolic" products that provide a sense of social status or "experiential" products that provide sensory/cognitive stimulation, while Mass brands are usually considered as "functional" products that provide some basic benefits to the customers. To position a brand based on the symbolic and experiential concepts, it is critical for the firms to develop specific distribution channel strategies to enhance the brand's image. For example, Channel may locate a specialty store at a five-star international hotel such as St. Regis or Ritz Carlton in Shanghai. Alternatively, a firm could sell its high-end products at some high-end Salon/Spa so that the sales representatives can provide personalized counseling services about skin-care or make-up to high-end customers in a more private environment. For instance, L'Oréal and Estée Lauder started their business ventures at various beauty salons in Paris and New York City in the 1900s and 1930s, respectively (c.f., Jones et al. (2006) and Koehn (2002)). Also, Paul Mitchell (http://www.paulmitchell.com) began his US$750 hair-care business venture in the 1980s at the beauty salons, which has grown into a US$700M success story in 2004. Besides distribution channel strategies, Premium brands are usually promoted as fashionable items that tend to have shorter product life cycles. As such, it is very difficult to obtain accurate demand forecast for the Premium brand products. Table 4 highlights different characteristics of Premium and Mass brand cosmetic products.

Table 4. Characteristics of premium and mass brands in China

Characteristic	Premium	Mass
Positioning	Symbolic or experiential	Functional
Competitive strategy	Differentiation (high perceived image)	Cost leadership (high perceived value)
Selling format	Counseling (high customer contact)	Self selection (low customer contact)
Customers	Upper income consumers (concentrated in major cities such as Beijing and Shanghai)	Less affluent city residents/affluent rural residents (scattered throughout China)
Profit margin	High	Low
Demand	Low volume/high mix (high demand uncertainties)	High volume/low mix (low demand uncertainties)
Product development cost	High	Low
Product/process technology	Complex	Simple
Product life cycle	Short (high obsolescence)	Long (low obsolescence)
Inventory cost	High	Low
Stockout cost	High	Low

4. Supply Chain Configurations

Based on our data collection and personal interviews in China, it appears that there are three major supply chain configurations that a multinational company would adopt as a supply chain strategy for its cosmetic products in China. We also noticed that some companies would use a combination of these configurations especially when they offer brands in multiple categories. The three major supply chain configurations are depicted in Fig. 3. Specifically, we have:

- *Direct or Indirect Export of Finished Products.* The firm produces the products in other regions such as Europe or North America and exports the finished goods to China directly to the distribution centers or indirectly through an agent or a distributor with local operating knowledge.
-
- *Outsourced Manufacturing.* The actual manufacturing function is outsourced to their Chinese suppliers under certain licensing or contractual agreements. However, the firm may export some or all of the raw materials to their Chinese suppliers including patented active ingredients such as Pitera. The basic raw materials such as Propylene glycol and Lanolin Alcohol in hand moisturizers, or packaging materials such as glass bottles and boxes are usually sourced within China.

- *Offshore Manufacturing.* This configuration is the same as outsourced manufacturing except that the manufacturing function is performed within a factory that is either a joint venture (JV) with a Chinese partner or a wholly owned subsidiary (WOS) of the firm. Besides serving the China market, the firm may ship the finished products to other Asian countries such as Singapore, Malaysia, India, Japan, and beyond. For example, Shiseido manufactures its Aupres JS products for men in Beijing and ships this line of products from China to the U.S.

Different Supply Chain configurations have different implications. We shall highlight three major implications for illustrative purposes. First, exporting cosmetic products to China can be beneficial. This is because Chinese consumers perceive the imported cosmetic products to be better in terms of brand image and quality. As such, to enhance the brand image, many multinational firms such as Estée Lauder continue to export their cosmetic products to China.

Second, outsourced manufacturing of cosmetic products can be risky. Even though the reform of the IP protection law has made some good progress after China's WTO entry in 2001 (http://www.chinaiprlaw.com/english/news/news5.htm), some unfortunate incidents can still occur in China. For example, as reported in Jones et al. (2005), Shiseido products were often counterfeited in China and Shiseido is currently working with

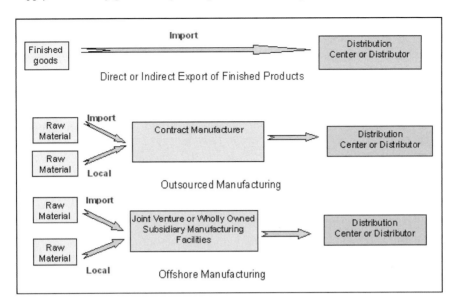

Figure 3. Three supply chain configurations

local police in Beijing to combat this problem. Also, multinational firms are not necessarily protected legally when their Chinese suppliers start producing unauthorized products using virtually identical design and materials. To elaborate, when the relationship between New Balance shoes and one of their Chinese suppliers went sour, this Chinese supplier started producing different types of shoes using a logo that resembles the New Balance's block "N" saddle design. New Balance filed a lawsuit in China without success and the saga continues. The reader is referred to Chandler and Fung (2006) for more details. As such, it is still difficult to protect IP and to eliminate the risk of counterfeits when a multinational firm outsources its manufacturing operations to its Chinese suppliers under certain licensing or contractual agreements.

Third, exporting cosmetics have other hidden costs. In addition to higher manufacturing and transportation costs, there are other hidden costs associated with tariff and time to market as well. For instance, the import tariff for various cosmetic products in China varied from 10 to 15% in 2004, but the tariff is supposed to reduce to 6% by 2008 under the WTO agreement (see http://www.china.org.cn/english/government/51861.htm). Moreover when exporting cosmetic products with certain claim of specific functions such as SPF, hair growth, blemish removal, etc., it takes usually 9 months to obtain special license from the Ministry of Health in China. In some cases, it would require inspection and quarantine, as imposed by the General Administration of Quality Supervision in China as well (c.f., Fang et al. (2006)). The process for getting proper license in China could delay the launch of certain specialty cosmetic products in China, which could affect the sales significantly. Table 5 provides a comparison among the three types of supply chain configurations.

Table 5. Characteristics of different supply chain configurations

Supply chain configuration	Direct/indirect export	Outsourced manufacturing	Offshore manufacturing
Perceived brand image	High	Low	Medium
Quality control	High	Low	Medium
Intellectual property protection	High	Low	Medium
Counterfeit risk	Low	High	Medium
Local market information access	Low	Medium	High
Time to market	Long	Short	Short
Sales volume potential[9]	Low	High	High
Startup cost[10]	Low	High	High
Manufacturing and transportation costs	High	Low	Medium
Responsiveness to demand change	Slow	Fast	Fast
Tariff[11]	High	Low	Low

By examining the characteristics of different brand categories in Table 4 and the characteristics of different supply chain configurations in Table 5, we establish the following hypotheses that examine the alignment between products and supply chain configurations:

- *To maintain a higher perceived image, multinational firms should export their premium brands to China.* While the time and cost associated with exporting could be higher, there are compensating factors including lower technology transfer risk, lower counterfeiting risk, economies of scale derived from producing the products at one location, better production capacity, and better control of quality and inventory.[12]

---

[9] The sales volume potential is even higher if one includes the potential of shipping the finished cosmetic products from the Chinese factories to other countries throughout Asia and beyond.

[10] Startup cost includes the cost of searching for a reliable contract manufacturer that could be significant and time consuming. This is exactly the core value provided by trading companies such as Li and Fung in Hong Kong. Specifically, Li and Fung provides a one-stop outsourcing solution to multinational firms such as Gap and Warner Bros. for the production of garments and toys (c.f., Tang (2006)).

[11] Besides import tariff, the Chinese government imposes import quotas for certain countries exporting certain products to China.

- *To offer a higher perceived value, multinational firms should either out-source or establish offshore operations in China for their mass brands.* While the financial risk and operational risk associated with this strategy are higher, there are compensating factors including lower manufacturing and transportation costs, strong brand awareness due to physical presence, better access to local market information, easier product localization to meet local consumer's preference, etc. Ultimately, having a stronger physical presence would enable multinational cosmetics firms to compete with domestic leaders such as Dabao in Beijing and Jahwa in Shanghai.

To test our hypotheses, we examine the supply chain configurations of different brands that are sold in China by gathering company information available in various research reports and by conducting company interviews in China. Our findings are summarized in Table 6.

Table 6. Supply chain configuration of different multinational brands in China

Brand category	Brand (company)	Export	Outsourced manufacturing	Offshore manufacturing
Premium	Lancôme (L'Oréal), Biotherm (L'Oréal), Estée Lauder (Estée Lauder), Clinique (Estée Lauder), Crème De Lar Mer (Estée Lauder), SK-II (P&G), Shiseido (Shiseido)	X		
Mass-premium	Revlon (Revlon), Aupres (Shiseido), L'Oréal Paris (L'Oréal)		X	X
Mass	Maybelline (L'Oréal), Olay (P&G), Dove (Unilever), Pond's (Unilever), Avon[13], Amway			X

---

[12] For instance, in early 1980s, due to a lack of sophisticated production technologies in China, Shiseido insisted that they can only produce basic toiletries in China such as shampoo and conditioner. These toiletries were sold in China under the brand name HuaZi without any association with Shiseido's name.

Our hypotheses are supported by the results reported in Table 5 especially for the premium and mass brands. First, to maintain a consistent global brand image focus, all Estée Lauder brands are premium products exported directly from Europe to China. Second, as Unilever produces only mass brand products, they need to compete with domestic leaders such as Dabao and Jahwa on cost and value. To do so, Unilever established joint ventures in Shanghai and Anhui. Also, for their mass brand products, P&G sources over 90% of their raw materials for their manufacturing production in China. Third, L'Oréal, P&G, and Shiseido pursued a diversification strategy in China by offering different categories of brands. As shown in Table 5, the premium and mass brands adopted the export model and the offshore manufacturing model, respectively, while the mass-premium brands adopted a hybrid model that combines outsourced manufacturing and offshore manufacturing. Specifically, in the hybrid model, the production of certain products associated with a brand is completely outsourced, while other products are produced under a mixed configuration (i.e., a mixture of offshore manufacturing and outsourced manufacturing). When a mixed configuration is adopted, the core operations that affect quality are usually performed at the offshore factories. These operations include the manufacturing of active ingredients, blending of different ingredients, and final assembly operations. Other non-core operations such as the manufacturing of packaging materials and basic ingredients are performed at various outsourced manufacturing sites.

5.   Distribution Channel Configurations

Based on our data collection and personal interviews in China, China's retail sector has undergone tremendous changes recently. Specifically, most cosmetic products are sold through the following five major distribution channels in China today:[14]

---

[13] According to the 1998 regulation established by the Ministry of Commerce of PRC, products sold within China under the direct sales model must be manufactured within China. Both Avon and Amway offer a few products in the Mass-premium category as well. To simplify our exposition, those mass-premium products are neglected here.

[14] Online sales of cosmetic products in China are still weak for three major reasons. First, most premium products sold online in China are counterfeited, smuggled, or bootleg products. Second, consumers are not fully protected for online frauds in China, even

- *Department Stores.* Department stores can be divided into two major classes: high-end stores such as Scitech in Beijing and low-end stores such as Beijing Xidan Shopping Center. In most cases, cosmetics establish their own sales counters at the department stores so as to provide professional advice to the customers. Since high-end department stores charge their rent based on a certain percentage of sales revenue (from 20 to 30%, depending on the location and the reputation of the department store), each cosmetics counter must satisfy certain brand image and sales record requirements.[15]

- *Hypermarkets/Supermarkets.* Major hypermarkets or supermarkets include Wal-Mart and Carrefour. Similar to department stores, cosmetics companies are required to pay certain "slotting fees" in order to sell their products at certain well known hypermarkets or supermarkets.

- *Specialty Stores/Professional Stores.*[16] Specialty stores tend to provide a full range of products of a specific brand and professional services such as skin test for skin care products or lighting simulation for color cosmetics. Well-known brands such as Shiseido and Estée Lauder set up their standalone specialty stores in China. Professional stores such as Sephora offer many premium brands in China since 2004.[17]

---

though all online sales websites are subject to government approval in China. Third, the logistics service industry in China is not well developed to support the pick-up and delivery services for online customers. For these reasons, few known brands offer online sales services directly to the customers in China. However, the customers in China can now purchase Avon products via Avon's authorized Chinese portal www.263.com; the products are shipped directly from Avon's regional distribution centers to the customers via 3rd party logistics providers (c.f., Gao et al. (2006)). Moreover, DHC of Japan has recently entered the China market using its website as its main distribution channel (c.f., Li and Fung Research Report, 2005).

[15] This practice is common among high-end department stores worldwide. For instance, without a proven brand image and sales record, Estée Lauder was unable to set up sales counters at Saks Fifth Avenue initially. However, Estée Lauder won Saks Fifth Avenue over after she devised an innovative scheme in the late 1940s (c.f., Koehn (2005)).

[16] Although there are no official figures on the sales at the beauty salons and spas in China, we observe many high-end beauty salons and spas are located in virtually all 5-star international hotels such as St. Regis, Four Seasons, Ritz Carlton, etc. These beauty salons and spas tend to attract high-end customers which can be a perfect location to test new products or to collect customer information in a more relax environment.

[17] Sephora (China) is a joint venture between Sephora of France and Shanghai Jahwa. It focuses on premium brands with Chinese characteristics. Sasa of Hong Kong also opened its first franchised cosmetic superstore in Shanghai in 2005 with a plan to open 30 stores by the end of 2007.

- *Pharmacies and Personal Care Stores.* Most pharmacies have pharmacists or medical doctors on site to provide professional advice to the customers. These value added services tend to attract customers who are seeking cosmetic products that offer specific functions such as wrinkle-free, whitening, and anti-acne, etc. For example, Vichy by L'Oréal was sold through pharmacies in China as a pharmacy brand (c.f., Fang et al. (2006)). Personal Care Stores such as Watson's have become increasingly popular for selling cosmetic products in China because of their credible and professional image.

- *Direct Sales.* Avon was the first cosmetic company introducing the direct sales method in China in 1990. Due to the concerns over the "pyramid selling" scheme, the Chinese government banned the direct sales method in 1998. The reader is referred to Godes (2004) for a detailed description of Avon's sales strategy and Coughlan (2004) for a detailed description of the pyramid selling scheme. To comply with the government regulation that each sales representative must be affiliated with a retail outlet, Avon has to set up its own standalone specialty stores and sales counters at department stores and supermarkets so as to sell its products. In April 2005, the Ministry of Commerce of China finally approved Avon to implement a pilot direct-selling program in Beijing, Tianjin and Guangdong. This pilot program is intended to help the Chinese government to develop ways to regulate the direct-sales channels properly. In 2006, Avon is selling their products via standalone specialty stores, cosmetic counters, direct sales representatives, and online via http://www.263.com, which has created some channel conflict issues for Avon to handle.[18] The reader is referred to Gao et al. (2006) and http://english.people.com.cn/200504/11/eng20050411_180456.html for more details regarding recent development of direct sales models in China.

Different Distribution Channels have different implications. First, as the beauty advisors at the department stores or specialty stores have in-depth knowledge about different products of a specific brand, these beauty advisors can provide professional advice that would enhance the perceived brand image as well as brand loyalty. In addition, they have specific customer knowledge, which is valuable to the company for developing new products, promotion plans, pricing strategies, etc. Second, the direct sales method is quite economical because the distribution cost is entirely variable.

---

[18] As reported in New York Times on April 29, 2006, Avon's first-quarter profit fell 67% due to the decline of direct sales in America; however, Avon is relying on the China market to sustain growth.

As such, it is easy for the company to expand without incurring upfront fixed cost such as slotting fee. Moreover, the direct sales method is an economical distribution channel for selling products in regions experiencing strong economic growth. These regions include Zhejiang, Shandong, Hubei, Henan, Sichuan, etc. Table 7 compares the key differences among different distribution channels.

Table 7. Characteristics of different distribution channels

Distribution channel	Depart-ment stores	Specialty stores/profes-sional stores	Hypermarkets/ supermarkets	Pharmacies and personal care stores	Direct sales
Perceived brand image	High	High	Medium/low	Medium/low	Medium/low
Sales volume potential	Medium	Medium	High	Medium	High
Distribution cost	High	High	Medium	Medium	Low
Product assortments	Medium	Large	Medium	Small	Large
Mutual learning	High	High	Low	Medium	High
Reach/convenience	Medium	Medium	Low (currently)	Medium	High
Relative change in market share	−33%	600%	+69%	100%	50%

Notice from Table 7 that the relative market share of cosmetics sold at the department stores has declined recently. For instance, as observed from Fig. 4, department stores accounted for 72% of all cosmetics sales in 1997; however, this figure has declined to 47% by 2004, which represents a decline of 33%. There are three major reasons for this trend. First, more channels especially hypermarkets and supermarkets are opening up more outlets in China. As such, some of the sales have shifted from department stores to other channels. Second, as customers are more familiar with certain brands or products, they are more comfortable to purchase the products at other channels besides department stores. Third, as Chinese government lifted the ban on direct selling, more sales representatives are eager to sell directly to the consumers in China. As more direct sales companies enter the China market, retaining experienced sales representatives is currently a major challenge that Avon and Amway are facing.

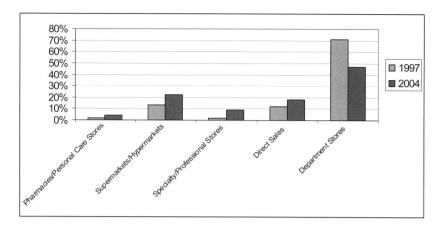

Figure 4. Changes of market share in different channels between 1997 and 2004 in China

By examining the characteristics of different brand categories in Table 4 and the characteristics of different distribution channels in Table 7, we establish the following hypotheses that examine the alignment between products and distribution channels:

- To maintain a higher perceived image, multinational firms should sell their premium brands via high-end department stores and specialty stores in China. While distribution cost associated with department stores and specialty stores are higher, there are compensating factors including brand image, customer intimacy, and consistent customer purchasing experience. These compensating factors could lead to customer loyalty.

- To offer a higher perceived value, multinational firms should sell their mass brands via hypermarkets/supermarkets and direct sales in China. While mass brands have lower profit margin, the direct sales and the supermarket channels would enable the firm to penetrate the consumer market in second-tier regions such as Zhejiang, Shandong, Hubei, Henan, Sichuan, etc. The additional sales generated from these second-tier regions in China could out weigh other negative factors.

To test our hypotheses, we examine the distribution channels of different brands that are sold in China by gathering company information available in various research reports and by conducting company interviews in China. Our findings are summarized in Table 8.

Our hypotheses are generally supported by the results reported in Table 8 especially for the premium and mass brands. For example, All Estée Lauder and L'Oréal's premium brands are sold through high-end department stores and specialty stores only. However, L'Oréal and P&G sell their respective mass brands Maybelline and Olay at various supermarkets, pharmacies and personal care stores, respectively. As L'Oréal expands its presence in every brand category and in every distribution channel, L'Oréal has the opportunity to learn more about the market conditions from the sales of different brand categories in different distribution channels.

Table 8. Distribution channels for different multinational brands in China

Brand category	Premium	Mass-premium	Mass
Brand (company)	Lancôme (L'Oréal), Biotherm (L'Oréal), Estée Lauder (Estée Lauder), Clinique (Estée Lauder), Crème De Lar Mer (Estée Lauder), SK-II (P&G), Shiseido (Shiseido)	Revlon (Revlon), Aupres (Shiseido), L'Oréal Paris (L'Oréal)	Maybelline (L'Oréal), Olay (P&G), Dove (Unilever), Pond's (Unilever), Avon, Amway
Department stores	X (High-end department stores only)	X (Mostly low-end department stores)	
Specialty stores/professional stores	X	X	
Hypermarkets/ supermarkets		X	X
Pharmacies and personal care stores		X	X
Direct sales			X

Combining the information provided in Tables 6 and 8, one can conclude that the multinational firms do align their supply chains and distribution channels to a great extent. Figure 5 illustrates how firms align their supply chains and distribution channels for different brand categories. For instance, most firms export their premium products to China and distribute them via high-end department stores and specialty stores. In addition, most firms manufacture their mass products at various offshore factories in China and distribute them at supermarkets and pharmacies or through direct sales representatives.

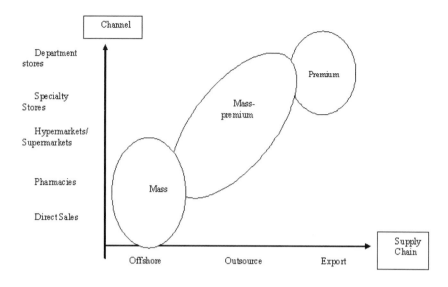

Figure 5. Supply chain and distribution channel alignment for different brand categories

6.   Outbound Distribution Configurations

There are two basic distribution configurations for companies to distribute their products in different channels. As depicted in Fig. 6, these two basic distribution configurations are:

- Direct Distribution. In this model, finished goods are shipped directly from the company's distribution center to the wholesalers or retailers.

- Indirect Distribution. In this model, finished goods are shipped to the key distributors or regional distributors first. After the title of the inventory is transferred to the distributors, the distributors or the regional distributors will then ship the products to different wholesalers or retailers.

Different outbound distribution configurations have different implications. We shall highlight two major implications for illustrative purposes. First, shipping directly to the wholesalers or retailers can be more economical especially when the foreign company focuses on a small number of whole-salers or retailers located in a few major cities such as Shanghai, Beijing and Guangzhou. Direct distribution enables foreign companies to postpone their shipments after receiving more accurate market information directly from the wholesalers and retailers. This postponement would shorten the replenishment

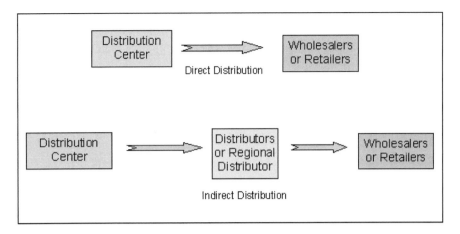

Figure 6. Two basic outbound distribution configurations

leadtime, which would enable foreign companies to mitigate the bullwhip effect and to make supply meet demand in a more efficient manner. The reader is referred to Lee et al. (1997) for details.

Second, shipping products via (regional) distributors have certain advantages including: lower financial risks because the distributors usually own the inventory; and stronger regional reach because the company only needs to ship to a few regional distributors instead of many outlets in different regions. Also, if a company distributes the products via multiple regional distributors, the company can switch regional distributors without major disruptions. However, indirect shipping has some drawbacks including: lower market visibility because the company does not deal with the wholesalers or retailers directly; lower profit margin because the distributors need to achieve a certain profit margin themselves; and lower sales volume because the distributors may increase the price they charge the wholesalers or retailers. Table 9 provides a basic comparison between these two outbound distribution configurations.

Table 9. Characteristics of different outbound distribution configurations

Outbound distribution	Direct ship	Indirect ship
Control	High	Low
Market visibility	High	Low
Inventory risks	High	Low
Gross margin	Higher	Lower
Geographical reach	Lower	Higher
Ease of market expansion	Relatively difficult	Relatively easy

By examining the characteristics of different brand categories in Table 4 and the characteristics of different outbound distribution configurations in Table 9, we establish the following hypotheses that examine the alignment between brand categories and outbound distribution configurations.

- To maintain a higher perceived image, multinational firms should ship their premium brands directly to the wholesalers or retailers in China. Since premium brands are usually sold in a small number of high-end retailers located in first tier cities, direct ship is more economical and it would enable the company to foster a stronger relationship with the retailers directly.

- To offer a higher perceived value, multinational firms should ship their mass-premium or mass brands indirectly via the distributors or regional distributors in China. As mass-medium and mass brands rely on sales in different regions, it is more economical for the company to ship the products indirectly via regional distributors especially because the regional distributors have local market knowledge.

To test our hypotheses, we examine the outbound distribution configure-tions of different brands that are sold in China. Except the fact that Avon and Amway ship directly to their sales counters and sales representatives via their own distribution centers, our hypotheses are supported by the results reported in Table 10.

Table 10. Outbound distribution configuration of different multinational brands in China

Brand category	Brand (company)	Direct ship	Indirect ship
Premium	Lancôme (L'Oréal), Biotherm (L'Oréal), Estée Lauder (Estée Lauder), Clinique (Estée Lauder), Crème De Lar Mer (Estée Lauder), SK-II (P&G), Shiseido (Shiseido)	X	
Mass-premium	Revlon (Revlon), Aupres (Shiseido), L'Oréal Paris (L'Oréal)		X
Mass	Maybelline (L'Oréal), Olay (P&G), Dove (Unilever), Pond's (Unilever)		X

7.   Research Opportunities

While our hypotheses are supported, there are many open research issues to be examined in the near future.

1. Optimal supply chain, outbound distribution, and distribution channel configuration. So far, we have examined the strategic alignment between supply chain, outbound distribution, distribution channels and brand categories. However, to truly operate an integrated (or coordinated) supply network, one needs to examine the following issues:

   * Supply Chain configuration: which manufacturing facilities, distribution centers, distributors, warehouses, wholesalers, and distribution channels should beselected?

   * Product assignment: which facilities (suppliers, manufacturing facilities, distribution centers distributors, warehouses, wholesalers, and distribution channels should be responsible for processing/selling which product so feach brand?

   * Customer assignment: which facility at an upstream stage should be responsible for handling the "demand" generated from downstream stages?

   * Production Planning: when and how much should each facility process?

   * Transportation planning: when and which mode of transportation should be used?

   Most of the published work in the area of supply network design is based on fixed and variable costs at each facility. For instance, Arntzen et al. (1995) and Camm et al. (1997) implemented a mixed integer programming model at Digital Equipment Corporation and P&G, respectively. Their models serve as planning systems for determining optimal decisions related to some of the above issues. However, there is no model that addresses the entire global supply network.

2. *Dynamic Supply Network Designs.* Most of the published work in the area of supply network design is based on a fixed network design. There is a need to develop new models for examining ways for companies to reconfigure their network designs so as to address market dynamics. This is an important issue especially when companies engage in the following activities:

* Change manufacturing facility locations. For example, Unilever has recently relocated its production facilities from Shanghai to Hefei in an attempt to lower the overall production cost.

* Increase the number of distributors or distribution centers. For example, P&G has added one additional distribution center in Xi'an recently.

* Increase the number of distribution channels. For example, Shiseido and Avon are expanding their distribution channels by opening specialty stores in major cities in China. Also, Shanghai Jahwa is opening professional stores via a joint venture with Sephora.

* Expand geographical regions. For example, L'Oréal exports its mass brand products from its factories in China to Japan, Korea, and other South east Asia countries.

* Acquire other brands. For example, as part of L'Oréal's global expansion strategy and as a way to fill the missing gap of mass and mass-premium brands in China, L'Oréal acquired two well-known domestic brands Mininurse and Yue-Sai in 2003 and 2004, respectively.[19]

We are not aware of any existing models for addressing the issue of dynamic configuration of a supply network.

3. *Product Rollovers.* To satisfy customer's variety seeking behavior, many cosmetics companies need to launch new products and phase out old products on an annual basis. It is challenging for companies to manage product rollovers especially when these changeovers affect the entire supply network including suppliers, manufacturers, distribution centers, distributors, wholesalers and retailers. There are very few models for guiding companies to manage the product rollover processes in an effectively manner. However, some basic framework and analysis have been developed recently. The reader is referred to Billington et al. (1998) and Lim and Tang (2006) for details.

4. *Emerging supply chain capabilities.* As more foreign cosmetics companies operating in China adopt more sophisticated information systems, there is a great opportunity for researchers to examine how these firms can utilize information technology to streamline their operations so as to

---

[19] L'Oréal's strategy focused on creating a strong presence in every brand category and in every distribution channel. This strategy has enable L'Oréal to achieve double digit growth in sales for the last 19 consecutive years (c.f., Jones et al. (2006)).

improve their overall supply chain performance.[20] For example, how should companies use point of sales data to develop more accurate forecasts? In addition, how should companies use information technology to increase supply chain visibility so that they can improve their coordination with their supply chain partners?

6. *Emerging logistics capabilities.* The logistics industry is fragmented due to regional license requirements in China. At this point, most supply chains in China are not integrated due to the need to use different transportation companies in different regions of China and due to limited IT capabilities to integrate information from different transportation companies. Currently, no single logistics provider commands more than 2% of the China market (c.f., Bolton and Wei (2005)). However, the logistics industry is growing by more than 50% annually since 2001 due to China's WTO entry, and it is changing rapidly due to recent mergers and acquisitions. For example, DHL (Germany) acquired Exel Logistics (England) in January 2006. As the logistics service operations become more sophisticated in China, many cosmetics firms can outsource their logistics operations to some major logistics providers so that they can focus on their core competence. More importantly, it would enable these companies to improve their supply chain operations so that they can offer more products to consumers without facing high operating cost, high inventory cost, or high obsolescence cost. Ultimately, it can enable these companies to compete on cost, product variety, and product availability (c.f., Billington et al. (1998) and Lee (2004)). There is a great opportunity for researchers to examine ways for foreign cosmetics companies to capitalize on the emerging logistics capabilities in China.

In summary, most multinational cosmetics firms align their supply chain configurations, distribution channels, and outbound distribution configurations to a great extent. However, firms with diversified portfolios in different brand categories tend to adopt multiple supply chain configurations and distributions channels. This strategy is consistent with the multiple supply chain concept articulated in Byrnes (2005). As the cosmetics market continues to grow in China, more research opportunities are present for managing supply chain reconfigurations, and ultimately, for managing company growth.

---

[20] P&G implemented an Efficient Consumer Response (ECR) system in China which has enabled them to improve customer service and reduce inventory.

References

Arnold, D., Strategies for Entering and Developing International Markets, Prentice Hall, 2003.

Arntzen, B., Brown, G., Harrison, T., and Trafton, L., "Global Supply Chain Management at Digital Equipment Corporation," Interfaces, Vol. 25, pp. 69–93, 1995.

Austin, J. and Aguila, F., "Nike in China," Harvard Business School Case # 9-386-065, 1985.

Billington, C., Lee, H., and Tang, C.S., "Product Rollover Strategies: Process, Strategies and Opportunities," Sloan Management Review, Vol. 39, No. 3, pp. 23–30, 1998.

Bolton, J. and Wei, Y., "Supply Chain Management in China: Trends, Risks and Recommendation," ASCET, pp. 66–68, Spring 2005.

Byrnes, J., "You Only Have One Supply Chain?" Harvard Business School Working Knowledge, Website: http://hbswk.hbs.edu/tools/, August 1, 2005.

Camm, J., Chorman, T., Dull, F., Evans, J., Sweeney, D., and Wegryn, G., "Blending OR/MS, Judgment, and GIS: Restructuring P&G's Supply Chain," Interfaces, Vol. 27, pp. 128–142, 1997.

Chandler, C. and Fung, A., "Not Exactly Counterfeit," Fortune, May 1, pp. 108–116, 2006.

Coughlan, A., "Mary Kay Inc.: Direct Selling and the Challenge of Online Channels," Kellogg School of Management Case # KEL 034, Northwestern University, 2004.

Ellis, M., "GM's Global Vehicle Sales Rise 4.4%," Detroit Free Press, April 20, 2006.

Euromonitor International Database, "Cosmetics and Toiletries in China," September 2005.

Fang, C., Gao, X., Liu, C.C., Wang, W.W., and Wu, T., "Analysis of Demand, Production, and Distribution of Cosmetics in Mainland China: Today and Tomorrow," Unpublished Report, UCLA Anderson School, 2006.

Forney, M., "How Nike Figured out China?" http://www.time.com, October 17, 2004.

Gao, X., Liu, C.C., Wang, W.W., Wu, T., and Xia, Y., "Analysis of Cosmetics Direct Sales Outbound Supply Chain: United States and China," Unpublished Report, UCLA Anderson, 2006.

Godes, D., "Avon.com (A)," Harvard Business School case # 9-503-016, Harvard Business School, 2004.

Jones, G., Kanno, A., and Egawa, M., "Making China Beautiful: Shiseido and the China Market," Harvard Business School case # 9-805-003, Harvard Business School, 2005.

Jones, G., Kiron, D., Dessain, V., and Sjoman, A., "L'Oreal and the Globalization of American Beauty," Harvard Business School case # 9-805-086, Harvard Business School, 2006.

Koehn, N., "Estee Lauder and the Market for Prestige Cosmetics," Harvard Business School case # 9-801-362, Harvard Business School, 2002.

Lee, H., "The Triple-A Supply Chain," Harvard Business Review, pp. 4–14, October 2004.

Lee, H.L., Padmanabhan, V., and Whang, S., "The Bullwhip Effect in Supply Chains," Sloan Management Review, Vol. 38, pp. 93–103, 1997.

Li and Fung Research Report, "The Booming Cosmetic Market in China," Li and Fung Group, Hong Kong, 2005.

Lim, W.S. and Tang, C.S., "Optimal Product Rollover Strategies," European Journal of Operational Research, Vol. 172, No. 3, pp. 956–970, 2006.

Pan, Y., "Whirlpool's Roadmap in China: 2004," Asia Case Research Center Report # HKU 414, Hong Kong University, 2005.

Porter, M., Competition in Global Industries, Harvard Business School Press, 1986.

Tang, C.S., "Robust Strategies for Mitigating Supply Chain Disruptions," International Journal of Logistics, pp. 33–45, March 2006.

Tao, Z., "China Cosmetics Industry 2005," Asia Case Research Center Report # HKU413, Hong Kong University, 2005.

Structural Supply Chain Collaboration Among Grocery Manufacturers

Timothy M. Laseter and Elliott N. Weiss
Darden Graduate Business School, University of Virginia, P.O. Box 6550, Charlottesville, VA 22906, USA

Abstract: Both academicians and practitioners have explored the important topic of supply chain collaboration, however, most have focused on unilateral coordination via pricing signals or tactical opportunities such as Collaborative Planning, Forecasting and Replenishment (CPFR). Our research fills a literature gap by quantifying the opportunity for strategic, multi-lateral collaboration through a shared distribution network. Through cost modeling of a factorial combination of scenarios for two grocery manufacturers, we show that:

1. Reducing demand variability, the focus of CPFR, lowers inventory levels but has little effect upon the larger elements of logistics cost: transportation and distribution center operations

2. Structural supply chain collaboration offers a large opportunity…but some forms of collaboration combined with certain manufacturer characteristics may add rather than reduce cost

3. The greatest opportunity for bottom-line savings accrues to the most disadvantaged manufacturers…those with low case sales, bulky products, and less costly product

1.   Introduction

Both academicians and practitioners have extensively explored the important topic of supply chain collaboration. Most research, however, has focused on the broad topic of strategic alliances or unilateral coordination via pricing signals. Practitioners, with some help from academics, have recently concentrated on the opportunities at the tactical level through Collaborative Planning, Forecasting and Replenishment (CPFR).

Our research fills a gap between these various streams by quantifying the opportunity for strategic, multi-lateral collaboration through a shared distribution network. Specifically, we examine the opportunity for manufacturers to consolidate their distribution networks to serve a set of common retailer customers as a first step towards consortia collaboration. Our analysis not only quantifies the savings potential, but also highlights the impact of key cost drivers. Though our initial analysis focuses on structural collaboration between two grocery products manufacturers, the concept could be extended to more participants as well as other industries.

The following section identifies the gaps in the current literature on supply chain collaboration. Section 3 explains our rationale and methodology for modeling a base case network of two hypothetical grocery manufacturers serving four common retailers. Section 4 examines different characteristics varied in the model to demonstrate the relative importance of structural supply chain collaboration for different manufacturers and presents the average results from three different collaboration scenarios. Section 5 examines two of the scenarios in detail to understand the relative importance of the manufacturing characteristics. Section 6 closes the paper with a discussion of managerial implications as well as the limitations of the current analysis and suggestions for future research.

2.   Supply Chain Collaboration Literature Review

Collaboration between supply chain partners has been covered extensively in the strategic management literature (Bradenburger and Nalebuff, 1996; Laseter, 1998; Spekman, 1990, 1992). Other researchers have examined the theoretical implications of supply chain collaboration through unilateral supply policies (Chen, 2001; Klastorin et al., 2002; Taylor, 2001, 2002; Wang and Gerchak, 2001). Other researchers have employed theoretical models to examine bilateral information exchange rather than unilateral policy incentives (Gavirneni, 2002; He et al., 2002; Li, 2002; Moinzadeh, 2002).

The practitioner-oriented research into supply chain collaboration tends to focus on the emerging concept of Collaborative Planning, Forecasting and Replenishment (CPFR) and Vendor Managed Inventory (VMI) (Aviv, 2001, 2002). Though rigorously reported empirical data are rare, (Lee et al., 1999) presents results for 31 grocery retail chains that adopted Campbell's continuous replenishment process and showed a significant improvement in inventory turns and a simultaneous reduction in stock-outs.

Sharing of physical assets as suggested by our concept of structural supply chain collaboration has been examined in pieces but not holistically. For example, Cheung and Lee (2002) and Granola and Chakravarty (2001) examine the benefit of coordinated shipments to improve truck utilization. Unfortunately, the concept of shared distribution networks – addressing the full range of logistics costs for inventory, transportation and warehousing – has not received adequate attention to date.

3. Base Case Distribution Scenario

Structural supply chain collaboration is a logical extension of current collaborative trends underway in a variety of industries. We anticipate, however, that the U.S. grocery industry may lead the way for three reasons. First, the U.S. grocery network offers a wide scope covering hundreds of factories and distribution centers serving an estimated 30,000 grocery stores. The distribution centers and full-truckload transport supporting this massive network are often outsourced to third party logistics providers (3PLs). These 3PLs have begun to experiment with shared facilities and truckloads to reduce some distribution redundancy... a natural first step towards true structural supply chain collaboration.

Second, supply chain collaboration ranks among the top priorities of leading consumer packaged goods manufacturers and their grocery retail partners. Their early pilots in Collaborative Forecasting, Planning and Replenishment (CPFR) delivered measurable results in improved product availability with simultaneously lowered inventory investments across the chain. Accordingly, the innovators are now exploring full scale implementations.

Third, the hype and/or threat at the peak of the Dot-com bubble triggered the creation of three powerful consortia models of B2B eMarketplaces: Transora, Worldwide Retail Exchange (WWRE) and Global Net Xchange (GNX). Founded and funded by the leading industry players, these eMarketplaces have helped to continue pushing the collaboration theme by standardizing software solutions among their members and encouraging further pilots.

Although these three factors suggest that the U.S. grocery industry may be ripe for structural supply chain collaboration, any movement in that direction will depend upon the bottom line economics. How much savings are possible? What type of companies will benefit the most?

As an initial step towards answering these questions, we analyzed a hypothetical grocery network structure comprised of two manufacturers and four retailers operating independently. From this base case network structure, we evaluated several scenarios as well as four critical variables in a factorial combination. Although a simplified and theoretical model, the analysis offers generalizable insights into the relative importance the individual variables.

As shown in Fig. 1, each manufacturer operates ten factories and six regional distribution centers and each retailer operates six distribution centers.

U.S. Grocery Network Model

For 2 Manufacturers and 4 Retailers

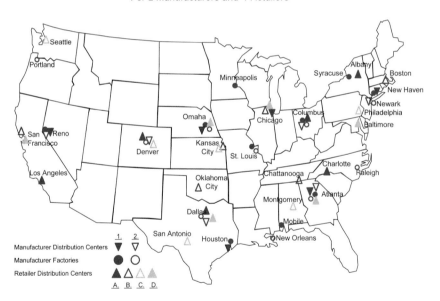

Figure 1. The distribution networks of the two manufacturers and four retailers have over-lapping locations but also important differences suggesting opportunities for collaboration

Although this base case network structure does not reflect the exact design for any particular manufacturer or retailer it nonetheless offers a reasonable reflection of actual industry practices. For example, Unilever is consolidating its national distribution network to six regional DCs – including one recently opened outside of Atlanta totaling over 1 million square feet (McCurry and Starner, 2002). The cities selected for the distribution centers in the base case include key ones commonly recommended for nationwide warehouse networks (Anonymous, 2002).

To explore the opportunities from structural supply chain collaboration, we built a financial model for a distribution network in the retail industry. Our cost model covers four categories of distribution cost for the manufacturer:

1. *Plant to Distribution Center (DC) Transportation*: the cost of tran-sportation from each factory to each manufacturer's distribution center
2. *DC to Retailer Transportation*: the cost of transportation from each manufacturer distribution center to the associated regional distribution center for the retailer
3. *Inventory Carrying Cost*: the carrying cost for inventory at the manu-facturer distribution centers based upon the lead time and variability for that site
4. *Distribution Center Operations*: the costs for leasing and operating the distribution centers – including an adjustment for economies of scale

To calculate the transportation costs for a full network, shipping costs are summed over every combination of origin and destination points. Between each dyad, the cost of a truck shipment is calculated based upon a fixed cost per trip and a variable cost per mile. The number of trips required between two points depends, in turn, on the total amount of goods to be shipped and the average utilization of the trucks. Multiplying the "per trip" value times the number of trips on the lane provides the total cost for the lane. Since a low utilization value can dramatically increase the actual costs (e.g., a 50% utilization rate doubles the number of shipments and accordingly doubles the cost relative to a full-utilization scenario), utilization plays a major role in driving the ultimate transportation cost. Combining volumes from two manufacturers should improve truck utilization due to the pooling of daily variation in shipments. We used a simulation model to estimate the utilization impact as a function of average lane volume and daily shipment variability. (For further explanation of the cost model and the transportation simulation, please contact the authors for a detailed supplementary document available under separate cover.)

Although the grocers benefit from increased truck utilization through the pooling of shipment variability, they do not get direct pooling effects in inventory. Since each sells unique products, the manufacturers must independently protect themselves from demand variability at the SKU level. Inventory levels will change, however, if the structure of the network changes due to collaboration. For example, a shared network with more distribution centers will reduce average pipeline inventory between the manufacturer DC and the retailer DC but will require more safety stock as the inventory is held in more locations.

To understand the impact of a shared network on distribution center operations we modeled the facility cost as a function of average inventory levels and throughput. For example, a distribution center located far from any factory holds more safety stock due to the longer replenishment lead times and accordingly needs more square footage. Other operating costs, such as picking and put away labor, are a function of throughput, independent of the average inventory level. We also applied a scale curve to reflect economies of scale from larger distribution centers versus smaller ones.

To populate the cost model with typical values we conducted primary research of public data sources such as U.S. government statistics, (Anonymous, 2001) published technical manuals (Ackerman, 1986; Digest, 1977) and textbooks (Arnold and Chapman, 2001; Harmon, 1993). Given that published sources tend to offer rough averages and become quickly dated, the most useful insight came from industry practitioners – manufacturers, logistics providers and consultants. A mix of sources allowed us to define reasonable values while avoiding disclosing any specific proprietary data from a specific grocery manufacturer.

To test the reasonableness of our model, we calculated the annual cost for operating the two networks independently without any form of collaboration and compared it to publicly reported data on the mix of distribution cost. As shown on the left hand side of Fig. 2, costs for our base case scenario total $413 million per year driven primarily by transportation cost. The mix of costs compares well to the split that exists for total logistics costs throughout the United States (Wilson and Delaney, 2001) shown on the right-hand side of Fig. 2.

The predominant role of transportation cost highlights that increased truck utilization likely offers the greatest opportunity for cost reduction. Less clear is the impact of the characteristics that define a manufacturer like its annual demand or the average price of its product. Will a cereal manufacturer like Kellogg's gain more from collaboration than a producer of dense products like Campbell's Soup? Section 4 examines a range of grocery manufacturer characteristics to develop insight into the relative opportunity for different consumer goods producers.

Logistics Cost Mix Comparison

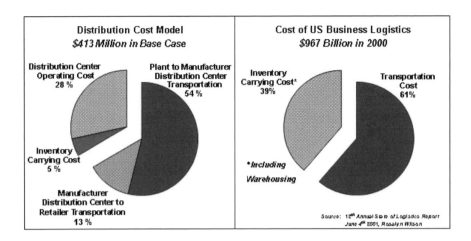

Figure 2. The cost model shows that transportation cost – particularly from plan to manufacturer DC – dominates overall distribution costs, consistent with US business in aggregate

4. Manufacturing Characteristics and Collaboration Scenarios

Grocery manufacturers have become increasingly global through a combination of organic growth and mergers and acquisitions. The leading grocery products manufacturers operating in North America are multibillion dollar businesses as shown in Table 1. However, the larger companies often must operate separate distribution networks to support different channels or product characteristics. For example, food products in permeable containers, like boxes of Lipton tea at Unilever could be contaminated if shipped and stored with Dove soap, another Unilever product.

Table 2 offers a further perspective on differences between grocery manufacturers. As can be seen, the number of cases per truckload – or Case Cube – varies from a low of 480 for cases of Cottenelle Toilet Paper to a high of 4,480 for Green Giant Asparagus Tips. An estimate of the cost per case, assuming a 25% gross margin at retail, range from a low of $5.99 for Bounty paper towels to a high of over $52 for Dove soap. Note that the values range substantially, even for Campbell Soup Company, due to different case packs for higher volume SKUs and the pricing difference for the more exotic soups.

2003 Manufacturer revenues

Table 1. North American revenues of major package goods companies range from $4 billion to $25 billion consistent with the hypothetical model

Grocery products Manufacturers	Total Revenues	Percent N. America	N. America Revenues
Campbell Soup Company	$6,133	71	$4,339
General Mills	$7,949	90	$7,139
H.J. Heinz	$9,431	55	$5,205
Johnson & Johnson	$33,004	61	$20,204
Kellogg	$8,853	69	$6,129
Kimberly–Clark	$14,524	64	$9,328
Kraft Foods	$33,875	74	$25,106
Nestle	$50,624	16	$8,100
Procter & Gamble[a]	$40,238	60	$24,143
Unilever	$45,914	21	$9,642

[a]*Percent North America estimated*

For the purposes of our modeling exercise, we identified four variables to capture the differences among manufacturers. *Annual volume*, the number of cases processed each year by the network, allows us to test the relevance of the shared network concept for manufacturers of different sizes. *Case cube*, the number of cases constituting a full truckload, helps us separate the producers of large, bulky items from the manufacturers of denser consumer products like canned goods. Likewise, *Case cost*, the wholesale price of the product, allows us to further differentiate for a given case density – for example, separating higher cost products like branded tuna fish from lower cost generic soup. Finally, we modeled *Daily plant variability* because reducing uncertainty and variability is typically a focus in most tactical efforts in supply chain collaboration.

For our base case we used 100 million annual case volume and $20 case cost suggesting North American revenues of $2 billion. In our full factorial model, we also examined scenarios which adjusted each variable up or down by 50%, producing a range for annual revenues as low $500 million (50 million cases at $10 per case) and as high as $4.5 billion (150 million cases at $30 per case). Our base case started with a case cube of 2,400 and 20% variability in the daily shipments from an individual factory. By similarly adjusting these two variables upwards by 50% and downwards by 50% we modeled 3^4 or 81 different combinations of manufacturer characteristics offering insight into the relative value of structural supply chain collaboration for different types of manufacturers.

Example Product Characteristics

Table 2. Case cube and case costs of a sample of leading products show the range possible and support the values used in the model

Product	Company	Size	Retail price ($)	Unit/ case	Cases/ layer	Layers/ pallet	Cases/ pallet	Pallets/ truck	Case cube	Cost/ case[a] ($)
Chicken noodle soup	Campbell soup	10.75 oz Can	0.99	48	7	5	35	40	1,400	35.64
Cream of shrimp soup	Campbell Soup	10.75 oz Can	1.29	24	7	10	70	40	2,800	23.22
Wesson oil	ConAgra	48 oz Bottle	2.99	8	12	3	36	40	1,440	17.94
Lucky charms cereal	General Mills	11.75 oz Box	3.99	12	9	4	36	40	1,440	35.91
Raisin, bran cereal	Kellogg	25.5 oz Box	4.39	14	6	3	18	40	720	46.10
Cottenelle, toilet paper	Kimberly–Clark	6 Rolls	3.29	8	6	2	12	40	480	19.74
Carnation, milk	Nestle	12 oz. Can	0.95	24	5	10	50	40	2,000	17.10
Green giant asparagus	Pillsbury	15 oz Can	2.69	12	14	8	112	40	4,480	24.21
Bounty paper, towels	Procter & Gamble	8 Rolls	7.99	1	10	4	40	40	1,600	5.99
Tide liquid detergent	Procter & Gamble	100 oz Bottle	8.29	4	10	3	30	60	1,800	24.87
Dove soap	Unilever	2 Bath Bars	2.89	24	15	5	75	40	3,000	52.02

[a]Assumes 25% gross margin at retail

In addition to examining different combinations of the four manufacturer variables, we also consider three alternatives to the base case (Scenario I) for structural supply chain collaboration. Scenario II considers the simplest and most obvious step towards collaboration between the two manufacturers: consolidating distribution centers in the two common cities of Atlanta and Reno. This step does not change the network footprint for either company, allowing them to continue operating with the same supply lead times and tactical process parameters. Accordingly, it requires no change in inventory levels. This initial scenario does, however, allow the two manufacturers to consolidate shipments from the four cities where the manufacturers both operate plants – Atlanta, Columbus, Omaha, and St. Louis – into the consolidated distribution centers in Reno and Atlanta. The manufacturers can also consolidate shipment from these two distribution centers to the Retailer DCs in the Southeast and the West Coast. The consolidation improves truck utilization along those lanes and thereby reduces transportation cost. This scenario also eliminates 2 of the 12 Distribution Centers in the network while expanding 2 of the remaining 10 to handle the increased volume. Because DCs exhibit some economies of scale, these larger DCs cost less to operate than the smaller ones they replace. Table 2 shows the relatively small shift in each of the cost elements as well as the total cost across the average of the combinations of the manufacturer characteristic variables (Fig. 3).

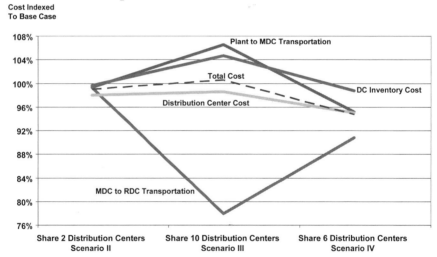

Average Costs of All 81 Combinations

Figure 3. An index of average costs of the elements making up total costs shows how the different scenarios create different cost tradeoffs

Scenario III offers a further level of integration beyond Scenario II – and produces a dramatically different cost picture. In this option, the two manufacturers maintain the same set of ten Distribution Centers, but reallocate volumes to share all of the facilities rather than just the two in

Atlanta and Reno. Sharing across the entire network increases inventory levels because each manufacturer now has safety stock in ten rather than six distribution centers which face greater uncertainty due to less pooling of the variability. In turn, distribution center costs increase from Scenario II as the DCs must be expanded slightly to accommodate the higher overall level of inventory in the system. Transportation cost from the factories to the manufacturer distribution centers also increases due to the fragmentation of demand across more lanes (i.e., each factory now ships to ten distribution centers rather than six thereby lowering the average daily volume per lane and accordingly the utilization rate). However, the new network requires far less transportation cost from the manufacturer DC to the retail DC because it allows for complete sharing in the shipments between the manufacturers and the shorter travel distance from a DC (a weighted average of 187 miles across the ten rather than 250 miles for the six).

Scenario IV requires the most significant structural collaboration and produces the lowest total cost on average. In this scenario, the two manufacturers consolidate their networks to a single shared distribution center in each of the six regions – effectively returning their networks to a similar structure as they operated individually but with the economies of scale in DC operations cost and the improved utilization of shared transportation. As would be typical in a real-life case, the decision to consolidate the networks opens up the opportunity to tweak and accordingly re-optimize versus the prior network structure as the design selected the best distribution center in each region as measured by the sum of the average inbound and outbound travel distances. Unlike the other cases, inventory drops in this scenario versus the base case: the number of inventory pools remains constant, but the slight reduction in inbound travel distance to the DCs reduces cycle, transit and safety stock for a slight savings. With the scale economies, Distribution Center costs drop by nearly 5% while simultaneously reducing the weighted average travel distances from factory to the manufacturer DC and on to the retailer DC.

A comparison of the individual cost elements across the three scenarios produces results that seem quite intuitive in retrospect. The inherent tradeoffs as well as the range of different values for the manufacturing characteristics, however, present an extremely complex set of relationships. As will be shown below, further insight comes from examining alternative ways of measuring the cost impact as well as the factorial combination of scenarios.

5.   Relative Savings Analysis

The preceding explanation demonstrates the general impact of the three structural collaboration scenarios. The characteristics of the manufacturers, however, impact the relative value of each scenario. Scenario II provides marginal but relatively consistent opportunity across all combinations of manufacturer characteristics. The savings range from a low of around six-tenths of a percent to 2.3% depending on the volume of truck traffic in the network. A manufacturer with the top level sales volume of 150 million cases annually and case cube at the low end value of 1,200 operates with nine times the volume of truck traffic versus one with 50 million cases annually and a case cube of 3,600 and accordingly faces less of a penalty from partially filled truckloads – and gains less benefit from sharing truckloads in a consolidated lane.

Scenarios III and IV produce a wider range of results. Sharing all ten Distribution Centers per Scenario III generates savings up to 2.6%...but can also cause costs to increase by 9.5%. The complete sharing of only six distribution centers, as captured by Scenario IV, produces consistently superior savings running from a low of 3.9% to a high of 9.4%. Figure 4 shows the logistics cost impact for all 81 combinations of the manufacturing characteristics for scenarios III and IV. Although an intricate chart with many data points and symbols, a detailed examination of it provides useful insight into the economic drivers of structural supply chain collaboration.

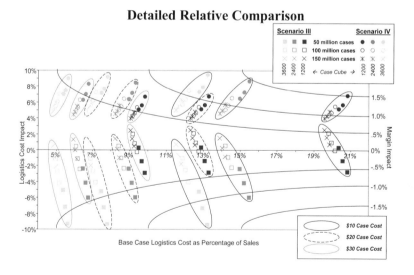

Figure 4. A plot of logistics savings and the margin impact shows the relative importance of different cost drivers

To aid in interpretation of the chart, different symbols are used for the three levels of annual case volume and different color/shading indicates different case cube levels. The symbols also distinguish Scenario III from Scenario IV as indicated in the legend box on the top right. Ovals encompass groups of nine data points with the same case cost level with the line style of the oval indicating the case cost as shown in the legend box on the bottom right. Demand variability is not indicated separately but displays a pattern of increasing base case cost as the value increases from 10–20% to 30% as indicated by the slight shifts to the right in the sets of three matching symbols within each oval throughout the chart.

The horizontal axis of the chart shows the logistics cost as a percentage of sales (assuming a 50% gross margin for the manufacturer) to help highlight the relative impact of logistics for the different combination of characteristics. Looking at the two ovals at the far right of the chart shows that the combination of a case cube of 1,200 and a case cost of $10 produces the highest logistics cost as a percent of sales...around 20 ± 1% depending on the annual volume and the demand variability. At the far left, the lowest percentage logistics cost – from 5 to 6% – occurs with the opposite combination: 3,600 case cube and a $30 cost per case.

The vertical axis on the left side of the chart indicates the impact of each scenario versus the base case which readily highlights that Scenario IV delivers consistently superior percentage reductions in logistics cost across all combinations. Scenario III offers a rough mirror image to Scenario IV: the greatest cost penalties for Scenario III occur with the same combination of characteristics that produce the highest savings in Scenario IV. The highest percentage cost savings for scenario IV and the biggest penalties for Scenario III occur with the combination of 50 million case volume, 3,600 case cube and 30% demand variability.

The scenarios produce the most similar results for large manufacturers shipping bulkier products with less variability as indicated by the combination of 150 million case volume, 1,200 case cube, and 10% daily demand variability. Although Scenario IV still dominates even for such a high shipping activity company, the difference is quite small and may be worth conceding to gain the advantage of faster deliveries to the retailer DCs. Scenario III offers ten distribution centers which reduces the transportation lead time to the retail distribution centers...a potential source of competitive advantage given increasing customer expectations regarding response times.

The vertical axis on the right side in combination with the curved lines indicates the impact on margins...a potentially more relevant consideration in assessing the financial impact of structural collaboration. An examination of the data points relative to the margin curves shows that although a high case cube and low annual volume produces the greatest percentage change in logistics cost, the greatest margin impact occurs with the low case cost of $10. This insight is particularly interesting in light of the observation that case cost has almost no affect on the percentage change in logistics costs since it only affects inventory cost – the smallest cost element in the total logistics cost equation. Though unimportant to the overall cost, case cost is critical to a manufacturer's perspective on the relative importance of the savings potential.

Through cost modeling of multiple combinations of the four variables, we have demonstrated that shared distribution can drive large savings for grocery product manufacturers. The annual case volume and the size of the cases (measured as "case cube") drives transportation cost which is the largest cost element in a distribution network. Although manufacturers of all sizes can benefit from structural supply chain collaboration, smaller volume manufacturers have the greatest percentage savings opportunity. This is particularly true for smaller manufacturers shipping expensive, bulky products with high variability. When looking at the saving potential in terms of margin impact, case cost becomes the most significant consideration.

6.   Conclusions, Limitations and Future Research

Obvious limitations of this paper stem from the fact that the model does not represent actual manufacturer networks and that some of the improvements identified result from better network optimization, independent of sharing. Also, the bottoms-up cost modeling approach employed does not take into account the issue of "transaction costs" as employed in the strategy literature (Coase, 1937; Commons, 1934). Furthermore, at this stage, the research also fails to address the potential implementation barriers to this new network design such as the governance model or change management issues.

Nonetheless this paper offers an initial perspective on an emerging concept of relevance to professional logisticians and managers at retailers, manufacturers, and third-party logistics providers (Drickhamer, 2003). By modeling four key variables for a hypothetical distribution network of two grocery manufacturers, the paper goes beyond theory to quantify the benefits and highlight the manufacturer characteristics likely to produce the greatest reward. This perspective is not currently highlighted in the supply chain collaboration literature.

Despite these limitations, the foregoing research and analysis highlights four important conclusions about the opportunity about supply chain collaboration:

1. Reducing demand variability, the focus of CPFR, lowers inventory levels but has little effect upon the larger elements of logistics cost: DC operations and transportation.
2. Structural supply chain collaboration offers a large opportunity...but some forms of collaboration may add cost rather than reduce cost for some manufacturers.
3. The greatest opportunity for bottom-line savings accrues to the most disadvantaged manufacturers...those with low case sales, bulky products, and less costly product.

The analysis also suggests that structural supply chain collaboration can provide a strategic opportunity for grocery manufacturers to partner with smaller retailers to offset the growing logistics scale advantage of the largest retailers such as Wal-Mart. Looking forward, we hope that other researchers will build upon the concept of structural supply chain collaboration and test its relevance for other supply chains and on other supply policies. Our results highlight the great potential for structural supply chain collaboration for the logistics community.

References

Ackerman, K. B. 1986. *Practical Handbook of Warehousing*, Second Edition: 104–105, 410–411, 457–458, Springer, New York.

Anonymous. 2001. *2000 National Occupation and Wage Estimates*, The Bureau of Labor Statistics, 2002.

Anonymous. 2002. *The 10 Best Warehouse Networks for 2002*. Chicago: Chicago Consulting1.

Arnold, T.J.R. and S.N. Chapman. 2001. *Introduction to Materials Management*. 340–359. Prentice Hall, New Jersey.

Aviv, Y. 2001. The effect of collaborative forecasting on supply chain performance. *Management Science* **47** (10) 1326–1343.

Aviv, Y. 2002. Gaining benefits from joint forecasting and replenishment processes: the case of auto-correlated demand. *Manufacturing & Service Operations Management* **4** (1) 56–74.

Bradenburger, A. and Nalebuff, B.J. 1996. *Co-operation: A Revolutionary Mindset that Redefines Competition and Cooperation in the Marketplace*, New York, NY: Doubleday.

Chen, F. 2001. Market segmentation, advanced demand information, and supply chain performance. *Manufacturing & Service Operations Management* **3** (1) 54–67.

Cheung, K.L. and Lee, H.L., 2002. The inventory benefit of shipment coordination and stock rebalancing in a supply chain. *Management Science* **48** (2) 300–306.

Coase, R.H. 1937. The nature of the firm. *Economic* **4** 386–405.

Commons, J.R. 1934. *Institutional Economics*, Madison, WI: University of Wisconsin Press.

Digest, E.o. D.W.C. 1977. *Digest of Warehouse Cost Calculations and Handling Standards*.

Drickhamer, D. 2003. Special delivery: logistics service providers want to manage you supply chain. Are they ready? *Industry Week* 24–27.

Gavirneni, S. 2002. Information flows in capacitated supply chains in fixed ordering costs. *Management Science* **48** (5) 644–651.

Granola, J. and Chakravarty, A., 2001. Sharing and lateral transshipment of inventory in a supply chain with expensive demand items. *Management Science* **47** (4) 579–594.

Harmon, R.L. 1993. *Reinventing the Warehouse*. 198, 243, The Free Press, New York.

He, Q.M., Jewkes, E.M., et al. 2002. The value of information used in inventory control of a make-to-order inventory-production system. *IIE Transactions* **34** (11) 999–1013.

Klastorin, T.D., Moinzadeh, K., et al. 2002. Coordinating orders in supply chains through price discounts. *IIE Transactions* **34** (8) 679–689.

Laseter, T.M. 1998. *Balanced Sourcing: Cooperation and Competition in Supplier Relationships*, San Francisco: Jossey-Bass Publishers.

Lee, H.G., Clark, T., et al. 1999. Research report. Can EDI benefit adopters? *Information Systems Research* **10** (2) 186–195.

Li, L. 2002. Information sharing in a supply chain with horizontal competition. *Management Science* **48** (9) 1196–1212.

McCurry, J. and Starner, R., 2002. *Unilever Unveils Expanded Southeast Distribution Center*, The Site Selection: Online Insider.

Moinzadeh, K. 2002. A multi-echelon inventory system with information exchange. *Management Science* **48** (3) 415–426.

Spekman, R.E. 1990. *Toward a Conceptual Understanding of the Antecedents of Strategic Alliances*. Cambridge, MA: Marketing Science Institute.

Spekman, R.E. 1992. *A Working Consensus to Collaborate: A Field Study of Manufacturer-Supplier Dyads*. Cambridge, MA: Marketing Sciences Institute.

Taylor, A.T. 2001. Channel coordination under price protection, midlife returns, and end-of-life returns in dynamic markets. *Management Science* **47** (9) 1220–1234.

Taylor, A.T. 2002. Supply chain coordination under channel rebates with sales effort effects. *Management Science* **48** (8) 992–1007.

Wang, Y. and Gerchak, Y., 2001. Supply chain coordination when demand is shelf-space dependent. *Manufacturing & Service Operations Management* **3** (1) 82–87.

Wilson, R. and Delaney, R.V., 2001. *12th Annual "State of Logistics Report." Managing Logistics in a Perfect Storm*. Washington, DC: Cass Information Systems, Inc. ProLogis: 24.

Supply Chain Management in the Chemical Industry: Trends, Issues, and Research Interests

Hong Choon Oh, I.A. Karimi, and R. Srinivasan
Department of Chemical and Biomolecular Engineering, National University of Singapore, Singapore

Abstract: Over the last decade, the global chemical industry has experienced one of its worst economic storms in recent history. In their bids to get through the unprecedented long economic downturn and to survive in the new economic era, many chemical companies have turned to reconfiguring their supply chains and revamping their operations. Inevitably, these structural and operational transformations and recent developments in the sociopolitical environment have brought several supply chain operational issues to the fore. This paper focuses on three primary issues. First, it identifies the key characteristics of supply chains in the chemical industry, which warrant management approaches that differ from those of other industries. Second, it discusses extensively the latest trends in chemical supply chains and the key supply chain operational issues that have unfolded over recent years in the chemical industry. The highlights of latest supply chain trends and issues enable readers to gain a good insight of current status as well as short term future of chemical supply chains. Third, this paper gives a brief overview of research work that has been done to address supply chain operation problems. Besides identifying gaps in the field of supply chain operation research, industrially relevant research opportunities that are gaining researchers' interests are also distinguished in this work.

1.   Introduction

Since the late nineties, the global chemical industry has sunk into economic doldrums. The woeful state of the chemical industry has also been augmented by a series of unfavorable events unfolded over the last few years that have crippling effects on the industry. The opening up the world's economy has led to competition of unprecedented intensity on the global stage. The last 5 years have witnessed flurry of chemical manufacturing activities migrating to low cost regions like Eastern Europe and Asia while the manufacturing bases in developed nations like United States and countries among Western Europe continue to shrink. Unfortunately, the increased in chemical plant capacities in low cost regions has taken its toll on the prices of their products as competitors fought fiercely to gain or retain their respective market shares. The economic downturn that hit the Asian region in 1997 and subsequently the economic powerhouse nations like US, Europe and Japan in early 2000s dampened the demand of chemicals. As a result, the downward pressure on the prices of chemicals is unusually prolonged.

As chemical companies struggled to cope with weak demands and declining prices of their products, the infamous September 11, 2001 terrorist attacks in New York threw almost all industries including the chemical industry into disarray. Chemical companies have to incur higher operating costs due to tightened security measures imposed by authorities to counter terrorism. The wars in Afghanistan (2001) and Iraq (2003) not only dashed hope of speedy recovery of the global economy. The retaliatory military actions led by the United States have also sent the feedstock prices soaring and further squeezed the margins of chemical companies (Glasgow, 2003). Overburdened with debts and poor economic outlook, the chemical industry's credit profile has slide to its worst level in memory in 2003 after a slew of credit downgrades (Chang, 2003) among the listed US chemical companies. This latest round of credit downgrades clearly bears testimony to the unfavorable sentiments of the global chemical industry. Moreover, the recent surge in crude oil price from about US$25 per barrel in April 2003 to more than US$60 per barrel in August 2005 may have brought cheers to the oil companies. But this spike in crude oil prices has also caused feedstock prices to skyrocket further which in turn has detrimental effect on margins of downstream chemical companies such as plastic makers. Due to intense business competition, many of these companies face difficulty in passing the feedstock increase to their customers, an issue which was highlighted in the 15th Asia Plastics Forum which was held in Singapore on April 25, 2005.

Despite being on their lows, companies in the chemical industry remain tenacious in their efforts to ride out of the economic turmoil. In their bids to last through the unprecedented long economic downturn and to survive in the new economic era, many chemical companies have turned to reconfiguring and revamping their supply chain designs and operations respectively. Inevitably, these structural and operational transformations in supply chain as well as recent developments in the socio-political environment have pushed several supply chain operational issues into the limelight.

The rest of the paper is organized as follows. We first discuss features of chemical supply chains which distinguished them from others before we describe the supply chain trends and issues that have unfolded over the years in the chemical industry. We then present an overview of past work that has been done to address the supply chain operation problems. Subsequently, we identify gaps and research areas in the field of supply chain operation researches that are gaining researchers' interests.

2. Unique Characteristics of Chemical Supply Chains

Chemical supply chains possess several characteristics which are distinctively different from those of supply chains in other industries. Clearly, understanding of these distinctive characteristics enables supply chain practitioners to appreciate the unique set of constraints and challenges that they have to contend. This is extremely crucial prior to the formulation and execution of any strategies that aim to manage chemical supply chain efficiently and effectively. Based on their areas of impact on supply chain decisions, we classify these distinguishing chemical supply chain characteristics into four main categories, namely material sourcing, manufacturing operation, demand and transportation management. For each of these categories, we now describe concisely the distinguishing characteristics of chemical supply chains.

2.1 Material Sourcing

Many chemical companies, including those in the oil and gas, petrochemicals businesses, usually source their raw materials in bulk. Moreover, many of these raw materials have been commoditized and are traded extensively in many exchanges around the world on a 24×7 basis. This is a sharp contrast compared to manufacturing companies of other industries where extensive commodity trading is virtually non-existent. As a result, opportunistic buying is often practiced in the chemical industry to exploit any significant cost saving opportunity. Hence, it is crucial that material sourcing decisions are made with good visibility of activities at trading exchanges as this ensures appropriate reaction is undertaken whenever a good trading deal arrives. But the option of exploiting any of such cost-saving opportunity must be exercised with caution as highly discounted raw materials may become highly discounted finished products when demand is a level that does not justify additional production.

Though many of the raw materials that chemical companies procure have been commoditized, variability in the qualities and compositions of these materials is an industry norm. Moreover, most chemical manufacturing processes entails product blending and multiple-recipes (to be discussed in greater details in the subsequent section) which inevitably make their outputs strongly dependent on the content of the raw materials used. Therefore, many material sourcing decisions have to be made with assistance of support tools that are able to evaluate usefulness of materials based on assay results and plant capabilities. Such tools are usually not employed in non-chemical industry because the latter consists of manufacturing processes that mostly do not involving product blending.

2.2   Manufacturing Operation

Many manufacturing processes of manufacturing plants essentially entail chemical reactions that are carried out in batch, continuous or semi-continuous operation modes with non-discrete products. They usually have multiple options of manufacturing recipes with complex nonlinear relationships between their raw materials and finished product, and several of these reactions even consist of multiple products being generated simultaneously. As such, numerous products and their variants of many chemical plants can be created from the same feedstock through blending of various constituents and the use of different process routes. Inevitably, production planning of their manufacturing processes has to contend raw material variability and product (including by-products) distribution issues which are usually addressed by feedstock blending and/or tweaking of process conditions and routes. Moreover, chemical plants usually store their non-discrete materials (raw materials, intermediate and finished products) in common storage tanks according to their identities or characteristics and not based on materials sources or product reaction pathways. Therefore, it is operationally impossible to link or tag each finished product to its corresponding raw material or process route. This limitation hinders root cause finding effort especially when product quality issues arise. On the other hand, majority of manufacturers from non-chemical industries do not have contend with this limitation since each of their manufacturing processes basically entails (1) production of discrete parts, (2) a fixed bill of materials (BOM), (3) single-product output, and (4) assembly-type processes.

Typically, chemical manufacturing facilities consist of complex networks of interconnected operating units for blending, separations, reactions and packaging. Operation of these facilities require tanks of various sizes to be setup within operating units and between units for temporary storage of raw materials, work-in-progress (WIP), and finished goods inventory. In addition, the immiscibility and incompatibility of the wide array of products used or produced in chemical plants (due to their properties) mean the different products can only be stored in different tanks that have different storage requirements. Process planning of chemical plants must recognize the limitations posed by real-time filling and emptying of all tanks in the system to avoid tank overflows and to respect cleaning requirements for product changeovers or maintenance. Inevitably, this makes production planning of manufacturing plants in chemical industry more complex than that in other industries since most of their manufacturing plants do not have to contend with complex constraints pertinent tank management.

Majority of the finished products of chemical plants serve as raw materials to manufacturing plants in chemical and other industries (i.e. most chemical companies conduct business-to-business (B2B) sales). In order to serve the needs of such wide variety of industries, most chemical plants produce in bulk and adopt make-to-stock approach. Therefore, they usually have to maintain higher inventories in their supply chain networks compared to non-chemical manufacturers. The latter, majority of which adopt make-to-order approach, have leaner inventory levels to meet demands of downstream users which primarily consists of distribution centers, retail outlets or individual end users.

All manufacturers distinguish their products based on their selected attributes. In non-chemical industries, these attributes are generally restricted to a limited set to tell apart different models, designs and model-specific options. However, attributes can assume an infinite range of values in chemical industry. This is because customers of chemical manufacturers usually specify their needs as "at least" or "no more than" a certain value of a given attribute. Thus, chemical manufacturers exploit this situation by substituting products of one quality (more or less of some attribute) with a product of higher quality when production efficiencies favor such "give away." Inevitably, production planning of chemical plants requires an understanding of product substitution and the rules of acceptable product replacements. Such requirement is usually not necessary among manufacturers from non-chemical industries.

2.3 Demand Management

As highlighted in Sect. 2.2, products of chemical plants can assume infinite possible range of attributes. Fortunately, customer orders are usually expressed in terms of "at least" or "no more than" certain value of a given attribute. Therefore, demand forecasting that chemical companies undertake not only have to be attribute-based, management of customer orders also require understanding of the underlying principle of substitution as well as the rules of acceptable product replacements as in production planning. In contrast, demand forecasting that non-chemical manufacturers undertake is based on their respective predetermined lists (i.e. finite number) of finished products which are differentiated by their designated store-keeping-units (SKUs). Essentially, no principle of substitution or rules of acceptable product replacements are required in order to manage their customer demands.

2.4   Transportation Management

Due to nature of their manufacturing operations, many chemical manufacturers have to coordinate their inbound and outbound trans-portation of materials (raw materials and finished products) in bulk. They employ a wide variety of transportation modes which include pipelines, tanker ships, tanker rail cars and tanker trucks to support the movement of their materials. The latter are usually hazardous in nature and their movement is usually governed by regulatory policies (that are legislated to address environmental, safety and security concerns) such as those imposed on the movement and tracking of hazardous materials. In addition, the immiscibility and incompatibility of these materials also mean that the transportation tools chosen to move them are subjected to maintenance requirements such as those pertinent to mandatory tank cleaning. In contrast, most manufacturers from non-chemical industry deal with raw materials and finished product that are chemically inert which are subjected to aforementioned regulatory or maintenance re-quirements. Moreover, their inbound and outbound transportation of materials are usually undertaken in volumes that are much smaller than those of their counterparts in the chemical industry. Evidently, transportation management of products across chemical supply chains is more complex than supply chains in non-chemical industry.

3.   Trend of Chemical Supply Chains

Over the years, several common trends are observed among chemical supply chains as chemical companies compete to survive in the new business landscape. Six of these trends are identified in this work and they are discussed in details in the following sections.

3.1 Proliferation of e-Commence

The phenomenal stride achieved in the field of information technology has led to increasing number of companies including those from chemical industry to make a paradigm shift in their business operations. Over the last 5 years, companies from practically all industries are scrambling to adopt the web-based technology to facilitate their business transactions within their organizations and with their respective business partners. The current stature of web-based technology has enabled each business transaction to be completed in just a few clicks of a mouse at practically any time of a day.

More companies are embracing the e-business approach because they recognize and value the benefits associated with such a move. The web-based approach of conducting businesses allows companies to reach out to a wider customer network base. It also improves the efficiency of moving products from suppliers to customers which in turn translates to cost-savings for both buyers and sellers. From a company's perspective, it is definitely a win-win situation for itself and its trading partners (suppliers and customers). An online buyer will be able to realize the best price for a given product since he/she has easy access of real-time market information. On the hand, a seller gains by clinching another online deal which contributes to the company's sales volumes.

Online auction site like ChemConnect® (http://www.chemconnect.com) is a typical example of where many chemical companies buy or sell their products in the new business era. ChemConnect® is a virtual trading floor where more than 9,000 companies from 150 countries trade their chemical products. Corporate chemical giants like Dupont, BASF, GE Plastics, Dow Chemical are among the companies that have used ChemConnect® as an avenue to trade their products. For example, Dow Chemical uses ChemConnect® to auction its off-grade plastics (acrylonitrile-butadiene styrene and styrene-acrylonitrile copolymer), engineering compounds and prime, off-grade and DVD grade polycarbonates. In 2002, transaction volumes of this auction site exceeded US$8.8 billion.

Though online trading of products offers an attractive and efficient means of conducting business, e-commence will never replace completely the traditional means of conducting businesses. This is especially true in certain sectors of the chemical industry where manufacturing of products like specialty-based chemicals (such as aroma chemicals) are tailored to the specific needs of customers. In such cases, the conventional practice of face-to-face negotiations, detailed specifications and qualifications remains the preferred channel of sealing business deals.

3.2 Waves of Restructuring

Since its last peak earnings year of 1995, the chemical industry has braved through waves of restructuring. The latter essentially takes the form of mergers and acquisitions (M&A), divestments and streamlining of business operations. Restructuring which involves overhaul of supply chains is undertaken primarily with the objective of improving a company's bottom line.

Examples of major M&A activities that have place in recent years include mergers of Exxon and Mobil, Chevron and Texaco, acquisitions of Aventis CropScience by Bayer, Dupont Textiles & Interiors by Koch Industries, Albright & Wilson by Rhodia and BTP by Clariant. The marriages of these chemical companies are mainly motivated by the opportunity of capitalizing on the potential cost synergies that accompanies any successful unification of companies. Nevertheless the realization of any cost saving cannot be achieved by M&A alone. The ensuing activities like strategic consolidation of plants, laying off of surplus human resources, divestments of non-core assets are crucial to the success of any M&A activities.

The current prolonged economic downturn faced by the chemical industry has confronted the chemical companies with sliding profits and downgrading of credit ratings. In order to reduce debt loads, many chemical companies have turned to divestment to improve their financial positions. Divestment is a powerful tool of value creation if it is approached strategically and executed professionally. It allows a company to get rid of its businesses which have poor growth prospects and may be a drag to future earnings. Sales of non-productive assets will not only improve a company's profit growth and return on equity. The cash proceeds from these sales also help to amortize the company financial debts. Though sales of non-productive assets are common in divestitures, some companies do divest their crown jewels in favor of a manageable balance sheet. For example, Crompton Corp sold its organosilicones business in 2003 to General Electric Company in a deal worth up to US$1.06 billion in cash, future payments and GE's plastic additives business. The organosilicones business is widely regarded a crown jewel of Crompton Corp and it's sale has helped the company to slash its debt of US$1.28 billion by more than 40%. See Table 1 for examples of pending divestments by chemical companies reported in 2004.

Table 1. Potential divestitures of chemical businesses reported in 2004

Company	Pending divestments
Akzo Nobel	Catalyst, resins and phosphorus chemicals businesses
Avecia	Specialty products division
Clariant	Electronic materials, textile dyes, custom synthesis, pharma chemicals and specialty fine chemicals businesses
MG Technology	Specialty chemical subsidiary, Dyanmit Nobel
RAG	Several businesses belong to subsidiary Rütgers, including its plastics processing business
Rhodia	Food and additives, phosphates and silicones businesses to raise €700 million

Source: Chemical Market Reporter (Chang, 2004a)

Over the years, the global chemical industry has also witnessed a flurry of M&A. Many chemical companies have turned to asset restructuring to improve their financial positions in the face of prolonged economic doldrums. In addition, the recent surge in M&A activity in the chemical industry is also fueled by the growing interests from financial buyers. After accounting for around 20% of chemical M&A activity annually from 2000 to 2002, financial buyers represented 38% of transactions in 2003. In dollar value, they also raked in 42% of total deals, compared to only 16% in 2002 (Chang, 2004a). This increase in M&A market share by financial buyers is mainly attributed a combination of factors that include low interest rates, the perception that the chemical industry is in its cyclical trough, and the conservatism shown by industrial buyers. Though the dollar volume of acquisitions of worldwide chemical companies appears to be falling in recent years, consolidation of chemical businesses are likely to remain active in the short term before the dollar volume returns to low level in early nineties when there were only US$5–6 billion in deals per year (see Fig. 1). It has been projected (Chang, 2004b) the acquisition deals in 2004 amounted to around US$30 billion.

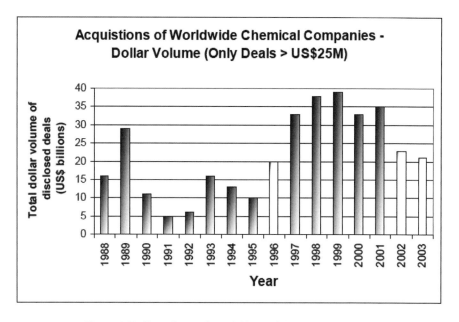

Figure 1. Dollar volume of acquisitions of chemical companies

3.3   Integration of IT Systems

The revolutionary innovations and developments accomplished in the information technology (IT) arena have led to many companies, including those from chemical industry recognizing the importance of supply chain software products as business tools for efficient business operations. Supply chain software applications are basically created with the goal of improving decision making processes across the entire supply chain via real time information sharing among different entities within an organization. The availability of real time information from different entities allows holistic and integrative approach in managing the entire supply chain of an organization. And this in turns creates an adaptive supply chain that is capable of responding swiftly to dynamics of today's markets. Timely shipments of finished products to customers, low logistics costs are, to name a few among of many benefits of deploying supply chain software tools. Over the years, the scope of supply chain software tools has also progressively broadened to include a company's suppliers and customers in the setup of the supply chain system infrastructures. The additional information tie up enables real time visibility of information (such as feedstock availabilities, feedstock prices and customer orders) from external entities which in turn can improve further the integrity of business decisions recommended by the software applications.

Modern supply chain software tools are generally marketed as suites of applications that manage different entities of a company's supply chain collectively. A growing number of these new tools, which are equipped with advanced programming or statistical routines, are developed with an open architecture that permits compatibility with other software systems regardless of their versions or suppliers. This open architecture concept of software development is especially important in the chemical industry where many companies have unenviable tasks of integrating various IT systems after the waves of M&A activities. In addition to meeting the rowing demands for open architecture software tools, technology vendors are also focusing on developing supply chain applications that conform to the ISA S95 guidelines that were issued by the International Institute of Measurement and Control in May 2000. Essentially ISA S95 establishes the common terms and definitions for types of information that exchanged between manufacturing operations and other business systems (like ERP). Table 2 summarizes some of the supply chain software tools that are commonly used in the chemical industry.

Ironically, many companies have committed a fundamental error in their rush to own the latest supply chain software applications. Many have failed to realize that integrative approach is crucial towards realizing the full value of such applications. As a result, heterogeneous networks of multiple decentralized IT systems have been evolved among many chemical companies. Consequently, the lack of information sharing among these multiple IT systems leads to automated business decisions being made based on local information instead of global one. The lack of integrative approach is exemplified by the installation of the Enterprise Resource Planning (ERP) systems by companies in chemical industry. ERP was created with the goal of integrating all departments and functions across a company onto a single computer system that can serve all those different departments' particular needs. But many big chemical companies decided to defy the basic working principle of ERP by customizing and decentralizing their ERP systems during the nineties. Despite being an early adopter of ERP, many companies in the chemical industry have yet to capitalize the full potential of an ERP system. On realizing that decentralized approach is a bad move, several chemical companies are now working on the laborious and time consuming task of integrating multiple fragmented systems into one. For example, Celanese recently embarked on a project of merging seven ERP systems into one (Berinato, 2003). This project, which was named OneSAP began in summer 2001 and was scheduled to be completed only in the mid-2004. The waves of M&A and divestment activities in recent and coming years are expected to result in more chemical companies focusing on consolidation or integrations of their ERP as well as other IT systems.

Table 2. Common supply chain software tools used in the chemical industry

Vendor	Tool name	Suite	Clients
Aspen Technology	Aspen MIMI™	• Aspen Strategic Analyzer • Aspen Collaborative Management • Aspen Supply Planner • Aspen Distribution Scheduler • Aspen Plant Scheduler • Aspen Capable to Promise	• Dupont • BP • Pharmacia & Upjohn
i2	i2 Supply Chain Management™	• i2 Demand Planner™ • i2 Supply Chain Planner™ • i2 Production Scheduler™ • i2 Demand Fulfillment™ • i2 Transportation Planner™ • i2 Transportation Manager™	• Bristol-Myers Squibb • Occidental Chemical Corporation (Oxychemical) • OPP Quimica
Manugistics	Manugistics Supply Chain Management	• Network Design and Optimization • Manufacturing Planning and Scheduling • Sales and Operations Planning • Collaborative VMI • CPFR™ • iHub • Service and Parts Management • Global Logistics Management • Global Logistics Sourcing • Fleet Management	• Petroiam Nasional Bhd (PETRONAS) • Mitsui Chemicals • Rohm and Hass Co.
SAP	mySAP SCM	• SAP Advanced Planning & Optimization • SAP Inventory Collaboration Hub • SAP Event Management	• BASF • Eastman Chemical • Dow Corning • MG Industries

3.4 Outsourcing of Non-Core Functions

As chemical companies are weathering the current tough economic climate, they are constantly on the lookout for cost saving measures to boost their bottom lines. One belt tightening measure that has prevalently been undertaken by chemical companies in recent years is the outsourcing of non-core business processes. Third party service providers are able to clinch these outsourcing business deals because they have the resources and expertise to run these outsourced processes more efficiently than their clients. More importantly, these third party service providers are also able to provide these efficient services at costs which are lower than those when the outsourced processes are managed by their clients. On the other hand, the handing off non-core business processes to the service providers allows chemical companies to channel their lean resources on their core operations. The types of business processes that have been commonly outsourced in the chemical industry are wide-ranging and are summarized in Table 3.

Table 3. Examples of outsourced functions of chemical companies

Category	Outsourced services
Human resource	Administration, recruitment, payroll processing
Finance	General ledger maintenance, fixed asset management, accounts payable processing
Logistics	Bulk breaking, drumming, shipments, storage, inventory, cleaning, waste management
Security	Health, safety, environment management
Information technology	Network infrastructure support, maintenance
Purchasing and sales	Procurement of raw materials, marketing and distribution of products
Others	Technology license management, manufacturing of intermediates

3.5 Growing Environmental Awareness

The International Council of Chemical Associations (ICCA) which represents the global chemical industry has been earnestly promoting Responsible Care program among its member countries. This voluntary program, a brainchild of Canada in 1985 has now been accepted by more than 46 countries where more than 85% (in volume) of the world chemicals are manufactured. Chemical companies which have enrolled into this program have to make conscientious effort to continuously engage in business operation and product improvement tasks that gear towards environmental, societal and economic sustainability. Though more than 46 countries have recognized the importance of Responsible Care, there is still variation in the level of participation by chemical companies in each of these nations. Some national associations like American Chemistry Council and Canadian Chemical Producers' Association have listed compliance to Responsible Care as prerequisite for memberships of their respective groups. But other national associations like the Singapore Chemical Industry Council do not have such restriction. The growing environmental consciousness among the chemical companies cum pressures from public and business partners are likely to result in more companies signing up the Responsible Care program voluntarily in the near future.

Besides the Responsible Care, several environmental initiatives have been meted out in recent years or with plans to roll over in the coming years. In 2003, European chemical producers and distributors have agreed on a set of guidelines on product stewardship throughout the chemical supply chains. Similar to Responsible Care, these guidelines are voluntary with the exception that they are extended to sectors allied to the chemical industry. The set of new guidelines stipulates requirements on handling, storage and transport, product and packaging disposal, packaging, classification and labeling of chemical products. And they are likely to provide a basis for legal contracts between suppliers and distributors in the chemical industry. The European community also has announced plans to implement several environmental legislations in the near future. The one which has direct impact on the chemical industry is the Restriction of Hazardous Substances (RoHS). The latter prohibits products containing lead, cadmium, hexavalent chromium, and other substances after 2006. In Asia, Japan has mandated in 2001 the requirement of manufacturers recycling lead-containing appliances and clearly document lead content of their products. Companies in United States also have to face a variety of recycling and take-back legislative initiatives.

In an era where protection of environment is steadily gaining its priority among organizations, chemical companies have to realigned their business operations and supply chain infrastructures to cope with the growing number of voluntary environmental guidelines or regulatory legislations. For example, chemical companies that manufacture lead-containing products may have to reconfigure their entire manufacturing processes to make their products lead free, or they may have to set up new infrastructure that tracks the lead-content of the products that are shipped to their customers. Chemical companies also may have to adjust their supply chain operations to manage the increase in delivery delays (from suppliers and to customers) that may arise due to the need to enforce the new environmental legislations.

3.6   Separation of Distribution and Logistics Functions

In Europe, the chemical supply chain is undergoing some infrastructural changes due to business need. Conglomerates that offer logistics and distribution services to chemical companies are beginning to offload their chemical distribution businesses in recent years. This trend which is expected to spread to other regions has arisen due to better money making potential of chemical logistics services compared to chemical distribution. The distribution service has suffered from poor margins in recent years due to the tough economic climate and its growth prospects are widely perceived to be poor in the industry. This has led to a relatively high number of chemical distributors being put up for sales in Europe. For example, Vopak spun off its Univar chemical distribution operation to focus on logistics operations in 2002. Deutscge Bahn AG (a German stateowned railway company) formally acquired Stinnes (a logistics conglomerate) in 2003 and has announced its intention to divest Brenntag which is formerly owned by Stinnes. Brenntag is the largest chemical distributor in Europe and Latin America and is among the top three in North America. Nevertheless, this separation of distribution and logistics functions is unlikely to have any major impact on the business operations of chemical companies.

4.   Current Chemical Supply Chain Management Issues

The new trends among the chemical supply chains as well as developments in the sociopolitical environment over the years have also pushed several supply chain issues into the limelight. Essentially, these issues are not new and many of them have existed for a long time. It is the new business environment which has forced several of these issues be among those that need to be addressed critically by supply chain practitioners.

4.1   Data Quality

Poor data quality can be attributed to three major causes, namely human error, systematic error and system incompatibility. Human errors are especially common among organizations where updating of databases is dominated by manual entries. On the other hand, systematic error arises when there is lack of standardization on the formats of information that is exchanged between entities (both internal and external) of a company's supply chain. This lack of standardization makes the exchanged information extremely prone to error. Lastly, system incompatibility usually arises when a company's IT infrastructure is equipped with multiple software systems that are not compatible with each other. Such awkward situation is especially common among companies in the chemical industry which have undergone waves of M&A activities in recent years. System incompatibility results untimely synchronization of data between the incompatible systems and this in turn contributes to poor data quality.

In a survey conducted by PriceWaterhouseCoopers (Moore and Gonzalez, 2002), it was revealed that only 15% of the respondents were "very confident" in the quality of data they receive from partners. And only 37% were "very confident" in their own data. This lack of confidence bears testimony to the extent of data quality issue faced by companies in almost all industries. Ensuring data quality is a time consuming and costly process. Nevertheless, companies have to acknowledge that their business operations remain inefficient and true value of their installed supply chain software tools cannot be realized till data quality is ensured.

4.2   Security Concerns

The unnerving security alerts that follow the terrorist attacks of September 11 have changed the business landscape of many industries. Scores of countries across the globe rushed to impose more-stringent custom checks and enact new security measures to counter terrorism. Consequently, companies are forced to revise their supply chain management practices within the constraints of the heightened security requirements. Additional contingency measures like higher safety inventory levels, enhanced information systems, expanded sourcing, storage and transportation bases are all becoming critical among companies in anticipation of any possible terrorist attack.

The global chemical industry is one of the many sectors that need to bear the blunt of economic blows that follow the September 11 attacks. Chemical companies not only have to contend with increasing operating costs due additional security and contingency measures to counter terrorism. They also have to cope with several proposals of new security legislations that threaten the cripple the slumping industry. For example, the Chemical Security Act was introduced in United States in less than 2 month after September 11. The act calls for chemical plants to access their vulnerabilities to terrorism and take steps to reduce them. It also pushes for reduction in usage and storage of hazardous chemicals by changing production methods and processes in the chemical plants. Wary of the massive costs needed for regulatory compliance, the American Chemistry Council (ACC) and the American Petroleum Institute lobbied successfully to prevent the bill from winning Congressional approval. Over in Europe, the European Commission proposed a new chemical testing regime called Registration, Evaluation,

Authorization of Chemicals (REACH). Though the proposal was not conceived as a result of terrorist attacks, it has similar goal as the Chemical Security Act which is to ensure high level of protection for human and environment by regulating the product testing systems in the chemical industry. The requirements of REACH system depend on properties, uses, exposure and volumes of chemicals produced and/or imported into the European Union. All chemicals with annual output of over 1 metric ton are to be registered in central database. However the REACH proposal was met with strong opposition from many chemical associations which include Japanese Chemical Industry Association, China Chamber of Commerce of Metals, Minerals and Chemicals Importers and Exporters, Philippines Chemical Industry Association and even European Chemical Industry Council. Many felt that REACH would be an unnecessary duplication of established Material Safety Data Sheets system and would adversely affect the global competitiveness of the industry. In response, the European Commission is currently working closely with all stakeholders to strike a balance between protection of its citizens' health, environment and workability of REACH for all companies involved. The High Level Group, which was tasked to conduct impact assessment of REACH only met recently on April 27, 2005 to discuss the findings of their impact studies.

The chemical supply chain security is currently receiving unprecedented level of attention from the public and the regulatory authorities. In the ensuing years, chemical companies will have to manage the unenviable task of balancing security and profits. And they have to do it in a daunting environment where any successful terrorist attack in the chemical industry is likely trigger a deluge of new uncompromising legislations to enforce security in this sector.

4.3   e-Business Risks

The evolution of e-business technology has offered a very lucrative alternative of making business transactions among companies. Though e-business has helped companies to improve their bottom lines, it has also brought upon additional risks that only the adopters of e-business technology have to contend with. Several ethical issues have relentlessly plagued the business world since the inception of information superhighway. Every company which undertakes online business activities is exposed to risks of hacking and fraud. Basically, computer hackers operate by searching for security flaws that allow them to have unauthorized access to companies' computer networks. Hackers capitalize on any successful intrusion by causing web outages, making fraudulent online transactions or even seeking ransoms from senior executives with threats of publicizing their intrusions.

Internet fraud has been widely identified across the global as one of the fastest growing and most pervasive forms of white collar crime. The multicultural and multinational nature of e-business has resulted in diverse views on how the cyberspace should be regulated and how cyber ethics should be enforced. For example, encryption of data transmission over the internet is an effective means of curbing e-business abuses. But many third world countries are still not in favor of this practice. On the other hand, investigation and prosecution of internet abuses usually require cooperation of law enforcement authorities from two or more nations. The lack of cooperation from any one of these nations will make it impossible to prosecute the cyber criminals who will in turn continue to pose threats in the e-business arena. Till the world reaches an agreement on how the cyber space should be regulated, companies with e-business activities from all industries have to contend with cyber ethic issues for many years to come.

5.   Research on Supply Chain Operation Problems

There is a wide variety of supply chain problems existing across every industry. One class of problems which has received extensive researchers' attention is the supply chain operation problems. The latter deal with operational aspects of supply chain management. These problems are evolved with objective of meeting strategic goals of a company in a given configuration of supply chain. Generally, supply chain operation problems involve business functional groups such as procurement department, production division and distribution department. These functional groups require sound planning to ensure smooth operation within each group as well as seamless integration across them.

Modeling these problems require extensive information from key components of supply chain to characterize these components individually and collectively. Depending on the approach and its underlying objective, research effort that has been undertaken to address supply chain operation problems can be classified into two main types. The first type of supply chain operation research entails process simulation of various supply chain entities to capture the supply chain dynamics. In general, supply chain simulation models involve mathematical representation of the characteristics of these individual supply chain entities as well as behavioral relationships among them.

This simulation approach is widely employed to formulate or analyze supply chain control policies such as inventory management policies, product sourcing and order allocation rules, etc. prior to the implementation of any policy. In most cases, supply chain simulation is done stochastically where samples are repetitively drawn from assumed distributions of uncertain parameters to generate distributions of selected performance measures.

The second type of supply chain operation research consists primarily of material flow planning within a given supply chain configuration and over short to medium term planning horizon. Typically, this requires application of optimization techniques to derive good (if not optimal) material sourcing, capacity allocation, product distribution plans that can meet customer orders and demand forecasts in the most economically efficient way. Depending on the gross margins of the businesses involved, there are two approaches by which the manufacturing processes can be represented in this field of research. For businesses with reasonable to high gross margins (e.g. consumer goods, specialty chemicals, pharmaceuticals), it is usually adequate to represent the manufacturing processes involved via recipe-based approach where operating conditions of the manufacturing processes are fixed and products are manufactured with fixed recipes. In contrast, it would be more appropriate to use the property-based approach to represent the manufacturing processes of businesses with slimmer margins (e.g. petrochemicals). In this approach, the operating conditions of the manufacturing processes are not fixed but subjected to bounds. The properties of each product stream depend on the operating conditions and underlying mixing rules.

5.1 Research Highlights

The supply chain operation problems have received wide spread attention from the operations research and chemical engineering communities for many years. As highlighted in Sect. 5, supply chain operation research essentially entails development of mathematical or simulation models that serve as decision support systems to address industrial supply chain operation problems. Progressively, the industrial realism of these models that have been developed to emulate real supply chain operation problems has improved significantly over the years. Most of the pioneering work is focused on simple problems such as those that only account for single manufacturing facility, single product, single echelon of distribution network or deterministic demand. Though the early resultant models lack industrial realism, they lay the foundations in their respective areas of supply chain issues and offer directions for future research activities. This is clearly demonstrated by the evolution of models that capture the complexity of real life supply chain operation problems over the years. Evidently, more dimensions in the form of multiple facilities, multiple products, multiple transportation options or multiple echelons distribution network have been integrated into supply chain operation models in recently published works than in the older papers. Such integrative models have evolved not only due to the need to improve the models' industrial realism but also to capitalize the benefits of approaching supply chain operation problems holistically. Moreover, supply chain intricacies like economies of scale in production or transportation costs, nonlinear manufacturing processes, local and international regulatory factors have increasingly been incorporated into supply chain operation models. The practical need to apply mathematical models to address industrial supply chain operation problems will continue to be the primary motivation for future improvement in model realism.

5.2   Research Opportunities

Over the years, it is apparent that significant progress has been made in the development of mathematical models to address supply chain operation problems. Nevertheless, there is still plenty of room for further improvement in this field of research. Powerful simulation frameworks like the multiagent based (e.g. Gjerdrum et al., 2000; Julka et al., 2002a, 2002b) and object-oriented (e.g. Hung et al., 2004) approaches which emerged only recently may have been applied in the development of supply chain simulation models. But these frameworks which consist of complex, stochastic, discrete event models contain adjustable parameters. To date, the application of optimization techniques to assign good values to these parameters has not been explored. It is also evident the competitive business environment has spawned several new property-based planning models (e.g. Jackson and Grossmann, 2003; Li et al., 2005) that are developed to support management of material flows across supply chains. With the stiff business competition likely to persist or even worsen, more models that are based on this approach are expected to be developed in the years to come to support industrial needs. In addition to the abovementioned, there are another four key areas that are gaining researchers' interests and that are industrially relevant to the chemical companies. We now describe each of them in more details as follows.

5.2.1 Account of Regulatory Factors

Among all papers from chemical engineering literatures that have been published to address supply chain operation problems, few have accounted for regulatory factors in their model constructions. As in Oh and Karimi (2004), these regulatory factors are referred to the legislative instruments (duties, tariffs, taxes, etc.) that a government agency imposes on the ownership, imports, exports, accounts, and earnings of business operators within its jurisdiction. The failure to incorporate international and domestic regulatory factors into supply chain models has virtually limited their application in the chemical industry. Therefore, it is extremely crucial to account all the relevant regulatory factors in the operation planning activity of any business. An optimal solution to a supply chain operation problem with a local focus will generally not be optimal, when one integrates several regulatory factors into the problem. Moreover, the global nature of business operations in chemical industry is accentuated by current competitive business environment where (1) scores of chemical companies across the globe are merging and streamlining their resources, (2) the suppliers, manufacturing plants, distribution centers and customers of most manufacturing firms are geographically spread around the world, and (3) trade agreements and disputes between countries or regions have perpetuated a heterogeneous network of trade barriers between countries in the world.

Though a fair amount of works by the operations research community has been done to address supply chain operation problems with international and domestic regulatory factors, most of these works do not account for operational features (e.g. multi-product manufacturing processes) that are unique to chemical companies. In addition, many of these works incorporate regulatory factors which are not accounted for in other works or vice versa. There is still lack of global supply chain operation models which comprehensively cover all the regulatory factors that have direct or indirect effect to any firm's bottom line. As far as research opportunities in supply chain operation problems of chemical industry are concerned, there is still abundance of them for researchers from both the operations research and chemical engineering communities to seize.

5.2.2 Stochastic Approach

All business operators have to contend with uncertainty over process parameters (e.g. yields, mixing rules), operational constraints (e.g. product availability, demands) and financial elements (e.g. prices, exchange rates, interest rates). Although extensive research has been done to address supply chain operation problems on a local scale with uncertainty being accounted, the uncertainty is usually assigned to one or two of the problem parameters. Full range of uncertainty that involve key process parameters, operational constraints and financial elements has yet to be explored. Moreover, there is still a dearth of stochastic approach on supply chain operation problems with international and domestic regulatory factors. Problems of such nature require knowledge and in-depth understanding of concepts from several fields. The latter include process engineering, international accounting and operations research. Process engineering knowledge is essential because understanding of how various components of supply chain are interrelated and knowledge of key processes' technical characteristics are needed to construct any realistic supply chain model. Knowledge of international accounting practices of multinational companies is crucial for understanding of constraints faced by multinational companies in managing their financial resources across their globally dispersed entities. Such knowledge is definitely critical for construction of realistic global supply chain operation model. And exposure to mathematical programming techniques that have been developed by operations research community to solve stochastic problems can certainly aid in the development of new solution methodologies to solve stochastic global supply chain operation problems. Apparently, the need to possess wide ranging scope of knowledge from various fields and the inherent complexity of stochastic global supply chain problems can pose a tremendous challenge to any researcher who ventures into this area of work. Nevertheless, any successful development of stochastic global supply chain operation model and a corresponding good solution methodology is likely to find its applications in the industry concerned with ease.

5.2.3 Disruption Management

Over the years, the world has been hit by a series of unexpected turbulent events that exposed the vulnerability of modern supply chains. The September 11 terrorist attacks in 2001, the labor strikes which cause West Coast port shutdown in 2002, the massive power outages that affected much of northeastern United States and Canada in 2003, the obliteration of oil refining and exploration facilities near the Gulf Coast by hurricane Katrina in 2005 are instances of turbulent events that have wrecked havoc to scores of supply chains. Many companies, which were ill prepared, have suffered heavy losses because their supply chains do not have the agility to respond effectively and efficiently to these disruptions. As a result, many multinational companies across practically all industries are beginning to look into ways of revising their supply chain configurations and practices so that they can operate in the event of serious disruption and in the most cost-effective manner. It is evident that this increased awareness of risks associated with supply chain disruptions has attracted interests from academics in this field of research in recent years. Development of supply chain operation models or frameworks that can serve as decision support tools in the presence of disruptions or to anticipate and prepare for disruptions is likely an emerging area that researchers may venture into.

6.   Conclusions

This global chemical industry is currently weathering one of its worst economic storms in recent history. In their bids to last through the unprecedented long economic downturn and to survive in the new economic era, many chemical companies have turned to reconfiguring and revamping their supply chains. This paper identifies distinguishing features of chemical supply chains which require management approaches that differ from those in supply chains of other industries. It also focuses on the latest trends of chemical supply chains that have evolved over the years. Moreover, three major supply chain management issues that modern chemical companies have to contend with are also discussed. Besides identifying gaps in the field of supply chain operation research, industrially relevant research opportunities that are gaining researchers' interests are also distinguished in this work.

References

Berinato, S., 2003, ERP consolidation: A day in the life of a big ERP rollup Celanese needs to merge seven SAP systems into one, a project expected to take 1000 days, *CIO*.

Chang, J., 2003, Chemical industry slammed with credit ratings downgrades, *Chemical Market Reporter*, August 18–25.

Chang, J., February 23 2004a, Chemical Mergers and Acquisitions Activity Heats Up as Financial Buyers Converge on Sector, *Chemical Market Reporter*.

Chang, J., November 15 2004b Chemical Industry M&A Activity Surges, *Chemical Market Reporter*.

Gjerdrum, A., Shah, N. and Papageorgiou, 2000, L. A combined optimization and agent-based approach for supply chain modeling and performance enhancement. *Production Planning and Control*, 12: 81–88.

Glasgow, B., March 3, 2003, Soaring feedstock prices crimp chemical industry profit margins, *Chemical Market Reporter*.

Hung, W., Kucherenko, S., Samsatli, N. and Shah, N., 2004, A flexible and generic approach to dynamic modelling of supply chains. *Journal of Operational Research Society*, 55: 801–813.

Jackson, J., Grossmann, I., 2003, Temporal decomposition scheme for nonlinear multisite production planning and distribution models. *Industrial and Engineering Chemistry Research*, 42: 3045–3055.

Julka, N., Srinivasan, R. and Karimi, I., 2002a, Agent-based supply chain management. 1: Framework. *Computers and Chemical Engineering*, 26:1755–1769.

Julka, N., Srinivasan, R. and Karimi, I., 2002b, Agent-based supply chain management. 2: A refinery application. *Computers and Chemical Engineering*, 26: 1771–1781.

Li, W., Hui, C. and Li A., 2005, Integrating CDU, FCC and product blending models into refinery planning. *Computers and Chemical Engineering*, 29: 2010–2028.

Moore, J. and Gonzalez, A., November 2002, Data quality management: The supply chain cornerstone, *Chemical Engineering Progress*, 29.

Oh, H-C. and Karimi, I., 2004, Regulatory factors and capacity-expansion planning in global chemical supply chains. *Industrial and Engineering Chemistry Research*, 43(13): 3365–3380.

Berth Allocation Planning Optimization in Container Terminals

Jam Dai, Wuqin Lin, Rajeeva Moorthy, and Chung-Piaw Teo

School of Industrial and Systems Engineering, Georgia Institute of Technology, Atlanta, GA 30332, USA

Department of Decision Sciences, National University of Singapore, Singapore 119260

Abstract: We study the problem of allocating berth space for vessels in container terminals, which is referred to as the berth allocation planning problem. We solve the static berth allocation planning problem as a rectangle packing problem with release time constraints, using a local search algorithm that employs the concept of sequence pair to define the neighborhood structure. We embed this approach in a real time scheduling system to address the berth allocation planning problem in a dynamic environment. We address the issues of vessel allocation to the terminal (thus affecting the overall berth utilization), choice of planning time window (how long to plan ahead in the dynamic environment), and the choice of objective used in the berthing algorithm (e.g., should we focus on minimizing vessels' waiting time or maximizing berth utilization?). In a moderate load setting, extensive simulation results show that the proposed berthing system is able to allocate space to most of the calling vessels upon arrival, with the majority of them allocated the preferred berthing location. In a heavy load setting, we need to balance the concerns of throughput with acceptable waiting time experienced by vessels. We show that, surprisingly, these can be handled by deliberately delaying berthing of vessels in order to achieve higher throughput in the berthing system.

1. Introduction

Competition among container ports continues to increase as the differentiation of hub ports and feeder ports progresses. Managers in many container terminals are trying to attract carriers by automating handling equipment, providing and speeding up various services, and furnishing the most current information on the flow of containers. At the same time, however, they are trying to reduce costs by utilizing resources efficiently, including human resources, berths, container yards, quay cranes, and various yard equipments.

Containers come into the terminals via ships. The majority of the containers are 20 feet and 40 feet in length. The quay cranes load the containers into prime movers (container trucks). The trucks then move them to the terminal yards. The yard cranes at the terminal yards then unload the containers from the prime movers and stack them neatly in the yards according to a stacking pattern and schedule. Prime movers enter the terminals to pick up the containers for distribution to distriparks and customers. The procedure is reversed for cargo leaving the port.

The problem studied in this paper is motivated by a berth allocation planning problem faced by a port operator. For each vessel calling at the terminal, the vessel turn-around time at the port can normally be calculated by examining historical statistics (number of containers handled for the vessel, the crane intensity allocated to the vessel, and historical crane rate). Vessel owners usually request a berthing time (called Berth-Time-Requested or BTR) in the terminal several days in advance, and are allowed to revise the BTR when the vessel is close to calling at the terminal. Note that the terminal operator allocates berth and quay side cranes according to the long term schedules and weekly ETA (Estimated-Time-of-Arrival) provided by the Lines. To minimize disruption and to facilitate planning, the terminal operator normally demands all containers bound for an incoming vessel to be ready in the terminal before the ETA. Similarly, customers (i.e., vessel owners) expect prompt berthing of their vessels upon arrival. This is particularly important for vessels from priority customers (called priority vessels hereon), who may have been guaranteed berth-on-arrival (i.e., within *two hours of arrival*) service in their contract with the terminal operator.

On the other hand, the port operator is also measured by her ability to utilize available resources (berth space, cranes, prime-movers, etc.) in the most efficient manner, with berth utilization,[21] berthing delays faced by

---

[21] This is the ratio of berth availability (hours of operations × total berth length) to berth occupancy (vessel time at berth × length occupied). The utilization rate is affected by the number and types of vessels allocated to the terminal. However, service performance (i.e.,

customers and terminal planning being prime concerns. We assume that the terminal is divided into several wharfs. Each wharf consists of several berths, which in turn are divided into sections. Each wharf corresponds to a linear stretch of space in the terminal. Figure 1 shows the layout of the four terminals in Singapore. Note that vessels can physically be moored across different sections, but they cannot be moored across different wharfs.

Our goal in this study is to design a berthing system to allocate berthing space to the vessels in real time. The objectives are to ensure that most vessels, if not all, can be berthed-on-arrival, and that most of the vessels will be allocated berthing space close to their preferred locations within the terminal. The allocation must be determined dynamically over time. To achieve this objective, we need to solve the following problem:

Static Berth Allocation Problem:

How do we design an efficient berth planning system to allocate berthing space to vessels in each planning epoch?

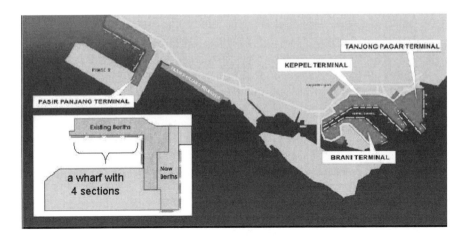

Figure 1. Layout of four terminals in Singapore

The above plan can then be embedded in a rolling horizon framework to allocate berthing space to vessels over time. However, while this may be able to meet the customers' request for high BOA service level, it may result in suboptimal use of the available terminal resources, in particular, low berth utilization. In a heavy load setting, this trade-off is especially crucial, since

on time berthing) will normally deteriorate with higher berth utilization, since more vessels will be competing for the use of the terminal.

the impact of inefficient resource utilization translates into a drastic drop in throughput. This leads us to the second issue related to the berth allocation planning problem:

Dynamic Berth Allocation Problem:

How do we incorporate features into the real time berth planning system to allocate berthing space to vessels, so as to o ensure on time berthing at the preferred locations for most of the vessels calling at the terminal, without adversely affecting the long term throughput?

Example: To see that such features are necessary and important components of a real time vessel berthing system, consider an example with three classes of vessels and a berth with seven sections.

- Each class 1 vessel occupies three sections. The arrival rate is one vessel every 16 h. The vessel turnaround time is 12 h.
- Each class 2 vessel occupies four sections. The arrival rate is also one vessel every 16 h. The vessel turnaround time is 14 h.
- Each class 3 vessel occupies five sections. The arrival rate is one vessel every 640 h. The vessel turnaround time is 16 h.

Suppose further that the nth class 1 vessel arrives at time $16n - 5$, class 2 at time $16(n - 1)$, and class 3 at time $640\,n$. Notice that, at most, two vessels can be processed each time. If we ignore the class 3 vessels, it is obvious that we can pack all the class 1 and class 2 vessels immediately upon arrival. The optimal packing is given by Fig. 2.

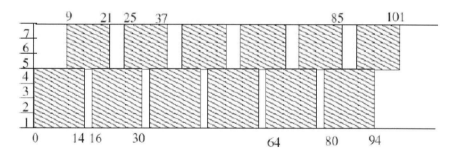

Figure 2. Optimal berthing schedule for class 1 and 2 vessels, when class 3 vessels are ignored

However, with class 3 vessels being assigned to the terminal, the berthing system must now address the trade-off between delays for vessels in the three different classes. How would a good dynamic berth planning system handle the load situation presented by this example?

If the focus is on minimizing total delays, then in each planning epoch, since each class 3 vessels will lead to delays on both class 1 and class 2 vessels, it is rational not to service class 3 vessels at all. This arises because class 3 vessels utilize five sections in the terminal (16 h for each class 3 vessel), leaving the other two sections unusable. However, the berthing plan obtained by ignoring the class 3 vessels will result in many small unused gaps within the berthing chart (cf. Fig. 2), resulting in lower utilization rate and lower throughput in the long haul (since class 3 vessels were ignored). Note that this low throughput scenario can be avoided if we use a berthing policy that ensures class 3 vessels can be moored immediately upon arrival, by reserving space in the terminal for class 3 vessels ahead of arrival. This, however, means that certain vessels from class 1 and 2 will have to be deliberately delayed even if the terminal has space available for them when they call at the terminal. This can be achieved with a priority differentiated berthing policy, or with a complicated penalty cost function to penalize against excessive delay experienced by the vessels. In general, how these can be incorporated in a dynamic berth allocation planning system and how these choices will complicate the static berth allocation planning problem, remains a challenging issue. In this paper, we borrow ideas and incorporate features from throughput optimal policies in the literature to address this issue. In this way, we ensure that the proposed dynamic berth allocation system will not lead to a situation where vessels will be delayed indefinitely in a heavy traffic environment.

In the rest of this paper, we outline an approach to address the issues associated with the static and dynamic berth allocation planning problem. Our contributions can be summarized as follows:

- We solve the static berth allocation planning problem as a rectangle packing problem with side constraints. The objective function chosen involves a delicate trade-off between the waiting time experienced and the deviation from preferred berthing space allocated. The optimization approach builds on the "sequence-pair" concept (cf. Imahori et al. (2003)) that has been used extensively in solving VLSI layout design problems. While this method has been proven useful for addressing classical rectangle packing problem to obtain tight packing, our rectangle packing problem is different as the key concern here is to pack the vessel efficiently using the available berth space. Using the concepts of "virtual wharf marks," we show how the sequence-pair approach can be augmented to address the rectangle packing problem arising from the berth allocation planning model.

- We extend the static berth allocation model to address the case where certain regions in the time-space network cannot be used to pack arriving vessels. These "forbidden" regions correspond to instances where certain vessels have already been moored and will leave the terminal some time later. In this case, the space allocated to these vessels cannot be used to moor other arriving vessels. The ability to address these side constraints allows our model to be embedded in a rolling horizon berth allocation planning model.

- In a moderate to heavy load setting, to ensure that throughput of the terminal will not be adversely affected due to the design of the berthing system, we propose a *discrete maximum pressure* policy to redistribute the load at the terminal at *periodic* intervals. We prove that this policy is throughput optimal even when processing times and inter-arrival times of vessels are random. Interestingly, this policy ensures that throughput rate attained is close to optimum, although the policy may deliberately insert delays on certain vessels, even when the terminal has enough resources to moor the affected vessels on time. In between the periods of load redistribution, however, the static allocation planning problem will be solved in each planning epoch to assign berthing space to the vessels. Hence this approach ensures that both the service performance of the berthing system (number of vessels berthed-on arrival, preferred berthing space allocated, etc.) and terminal throughput performance are addressed by a single berthing system.

- We conclude our study with an extensive simulation of the proposed approach, using a set of arrival patterns extracted and suitably modified from real data. As a side product, our simulation also illustrates the importance of the choice of planning horizon, and the trade-off between frequent revision to berthing plans and berthing performance. Note that frequent revision of plans is undesirable from a port operation perspective, since it has an unintended impact on personnel and resource schedules.

2.   Literature Review

There is by now a huge literature on the applications of operations research in container operations (cf. Steenken et al. (2004)). To the best of our knowledge, while there are sporadic papers that address the static berth allocation planning problem, none of the papers in the literature addresses specifically the issue of throughput loss in the dynamic berth planning allocation model. Brown et al. (1994) formulated an integer-programming model for assigning one possible berthing location to a vessel considering various practical constraints. However, they considered a berth as a collection of discrete berthing locations, and their model is more apt for berthing vessels in a naval port, where berth shifting of moored vessels is allowed. Lim (1998) addressed the berth planning problem by keeping the berthing time fixed while trying to decide the berthing locations. The berth was considered to be a continuous space rather than a collection of discrete locations. He proposed a heuristic method for determining berthing locations of vessels by utilizing a graphical representation for the problem. Chen and Hsieh (1999) proposed an alternative network-flow heuristic by considering the time-space network model. Tong et al. (1999) solved the ship berthing problem using the Ants Colony Optimization approach, but they focused on minimizing the wharf length required while assuming the berthing time as given. In the Berth Allocation Planning System (BAPS) (and its later incarnation iBAPS) developed by the Resource Allocation and Scheduling (RAS) group of the National University of Singapore (NUS), a two-stage strategy is adopted to solve the BAP problem. In the first stage, vessels are partitioned into sections of the port without specifying the exact berth location. The second stage determines specific berthing locations for vessels and packs the vessels within their assigned sections. See Chen (1998) and Loh (1996). Moon (2000), in an unpublished thesis, formulated an integer linear program for the problem and solved it using LINDO package. The computational time of LINDO increased rapidly when the number of vessels became higher than seven and the length of the planning horizon exceeded 72 h. Some properties of the optimal solution were investigated. Based on these properties, a heuristic algorithm for the berth planning problem was suggested. The performance of the heuristic algorithm was compared with that of the optimization technique. The heuristic works well on many randomly generated instances, but performs badly in several cases when the penalty for delay is substantial. For a follow up paper see Kim and Moon (2003). Park and Kim (2003) use a Lagrangian relaxation based approach to solve the berth planning problem. Assuming a convex cost structure, they also describe a heuristic procedure for obtaining feasible solutions from the solutions to the relaxed problem. Though relevant, the approach is limited by the choice of the cost function. Further, they do not describe how the methodology can be incorporated in a dynamic framework.

The static berth planning problem is related to a variant of the two-dimensional rectangle packing problem, where the objective is to place a set of rectangles in the plane without overlap so that a given cost function will be minimized. This problem arises frequently in VLSI design and resource-constrained project scheduling. Furthermore, in certain packing problems, the rectangles are allowed to rotate 90°. In these problems, the objectives are normally to minimize the height or area used in the packing solution. Imahori et al. (2003), building on a long series of work by Murata et al. (1996), Tang et al. (2000) etc., propose a local search method for this problem, using an encoding scheme called *sequence pair*. Their approach is able to address the rectangle packing problem with spatial cost function of the type $g\left(\max_i p_i\left(x_i\right), \max_i q_i\left(t_i\right)\right)$, where p_i, q_i are general cost functions and can be discontinuous or nonlinear, and g is nondecreasing in its parameters. Given a fixed sequence pair, Imahori et al. (2003) showed that the associated optimization problem can be solved efficiently using a Dynamic Programming framework. Unfortunately, the general cost function considered in their paper cannot be readily extended to incorporate the objective function considered in this paper.

The berth allocation planning problem is also related to a class of multiple machines stochastic scheduling problems on jobs with release dates, where each job needs to be processed on multiple processors at the same time, i.e., there is a given pre-specified dedicated subset of processors which are required to process the task simultaneously. This is known as the multiprocessor task scheduling problem. Li et al. (1998) focused on the makespan objective and derived several approximation bounds for a variant of the first-fit decreasing heuristic. However, their model ignored the arrival time aspect of the ships and is not directly applicable to the berthing problem in practice. In a follow-up paper, Guan et al. (2002) addressed the case with weighted completion time objective function. They developed a heuristic and performed worst case analysis under the assumption that larger vessels have longer port stays. This assumption, while generally true, does not hold in all instances of container terminal berthing. Guan et al. (2004) consider two mathematical formulations for berth allocation with the weighted flow time objective and propose a heuristic procedure for solving large size problems. They group vessels into batches and assume that vessels within a batch arrive at the same time. Their approach is tactical in nature and cannot be adapted to schedule vessels in a dynamic setting. Fishkin et al. (2001), using some sophisticated approximation techniques (combination of scaling, LP, and DP), derived the first polynomial time approximation scheme for the multi-processor task scheduling problem, for the minimum *weighted completion time* objective function. However, in the berth planning problem, since release time of each job is given, a more appropriate objective function is to consider the weighted *flow time* objective function. Unfortunately, there are very few approximation results known for scheduling problems with mean flow time objective function.

The closest literature to our line of work is arguably the series of papers by Imai et al. (2001, 2003). In the first paper, they proposed a planning model where vessel arrival time is modeled explicitly and they proposed a Lagrangian-based heuristic to solve the integer programming model. Their model is a simplified version of our static berth allocation planning problem, since they implicitly assume that each vessel occupies exactly one berth in their model. The issue of re-planning is also not addressed in that paper. In the second paper, they proposed an integer programming model and a genetic algorithm for solving the static berth planning model with differentiated priorities, but their model assumes deterministic data and does not take into consideration the arrival time information.

3.   Static Berth Allocation Planning Problem

At each planning epoch, given information on the vessels that will be arriving within the chosen planning horizon (or scheduling window), we structured the associated berth planning problem as one of packing rectangles (without overlap) in a semi-infinite strip with general spatial and time cost structure. In this phase, the packing algorithm must minimize the delays faced by vessels, with higher priority vessels receiving the promised level of services. At the same time, the algorithm must also address the desirability, from a port operator's perspective, to moor the vessels on designated locations along the terminal and to minimize the movement and exchange of containers within the yards and between vessels. These preferred locations for the vessels are normally pre-determined based on the containers storage locations and transshipment cargo volume of the vessels.

The inputs to the problem are:

- Terminal specifics:
    - W: Number of wharfs in the terminal,
    - M: Number of berths in the terminal,
    - b_l: Length of berth l, l = 1, ..., M.

- Vessel specifics:
    - N: Number of vessels that have arrived or will arrive within the scheduling window,
    - r_i: Arrival time (ETA) of vessel i,
    - l_i: Length-overall of vessel i,
    - p_i: Length of port stay (turnaround time) upon berthing by vessel i.

The parameters l_i and p_i specifies the dimensions of the N rectangles that need to be packed within the semi-infinite strip, and r_i represents the release time constraint imposed on vessel *i*. For ease of exposition, we assume that vessels can be moored at any berth available within the terminal. In real terminal berth allocation planning, we normally have to take into account other issues like draft constraints, equipment availability along the berth, tidal information and crane availability etc., to determine the berth where the vessels will be moored. The approach proposed herein can be easily extended to address a host of such side constraints. We also note that although we take the vessel-specific parameters as given constants for each scheduling window, in reality, some of the parameters (such as arrival time within the scheduling window) may deviate from forecast. In particular, the port stay time p_i depends on crane intensity (average number of cranes working on the vessel per hour) assigned to vessel *i*, and also on the number of containers to be loaded and discharged. Nevertheless, the impact of such inaccuracies can be minimized by re-optimization in a rolling horizon framework.

The decision variables to the berth planning problem are:

- x_i: Berthing location for vessel i, measured with respect to the lower left corner of the vessel, represented as rectangle in the time-space plane,
- t_i: Planned berthing time for vessel *i*.

Given a berthing plan with prescribed decisions x_i and t_i, we can evaluate the quality of the decision by two cost components:

- **Location Cost**: The quality of the berthing locations assigned is given by

$$\sum_{i=1}^{N} c_i(x_i)$$

where the space cost function $c_i(\bullet)$ is a *step function* in x_i defined as,

$$c_i(x_i) = \begin{cases} d_{i,1} & \text{if} \quad x_i \le b_1, \\ d_{i,2} & \text{if} \quad b_1 < x_i \le b_1 + b_2, \\ \vdots & \vdots \quad \vdots \\ d_{i,M} & \text{if} \quad \sum_{j=1}^{M-1} b_j < x_i \le \sum_{j=1}^{M} b_j. \end{cases}$$

Note that $d_{i,l}$ indicates the desirability of mooring vessel *i* in berth *l*. This depends on where the loading containers are stored in the terminal and also on the destination of the discharging containers.

- **Delay Cost:** The quality of the berthing times assigned is given by

$$\omega_i \times f_i(t_i)$$

where, the function $f_i(t_i)$ is a non decreasing function of t_i. Usually, a vessel is considered "berthed-on-arrival" (BOA) if it can be moored within 2 h of arrival (i.e., within r_i and $r_i + 2$) and hence, the delay cost function is typically defined to be $f_i(t_i) = (t_i - r_i - 2)^+$. The value ω_i is a penalty factor attached to vessel k and indicates the "importance" of berthing the vessel k on arrival (i.e., within a 2-h window). The vessels are normally divided into several priority classes with different penalty factors. Note that the approach proposed in this paper works for all non-decreasing cost function *f*.

The berthing decisions x_i and t_i must satisfy a host of other constraints. Most notably, the choice of x_i and t_i must not lead to any conflict in terminal space allocation. i.e., no two vessels can be assigned to the same stretch of space in the terminal at the same time. Furthermore, since the vessels cannot be moored across different wharfs, x_i is further restricted to belong to disjoint segments of space within the terminal. Coupled with the release time constraints, the static berth allocation planning problem belongs to a class of non-convex optimization problem which is exceedingly difficult to solve. In the rest of this section, we model this problem as rectangle packing problem and use a special encoding scheme to reduce the optimization problem to a search over pairs of permutations.

3.1 Sequence Pair Concept

We first show that every berthing plan can be encoded by a pair of permutations (H, V) of all vessels. Consider the berthing plan of two vessels as shown in Fig. 3.

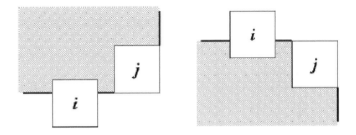

Figure 3. The figure on the *left* shows the LEFT-UP view of vessel j, whereas the figure on the *right* shows the LEFT-DOWN view of vessel j in the time–space plane; the views seen by vessel views seen by vessel j are *shaded* regions plus the parts hidden by vessel i

The permutations H and V associated with the berthing plan are constructed with the following properties:

- If vessel j is on the right of vessel i in H, then vessel i *does not* "see" vessel j on its LEFT-UP view.
- Similarly, if vessel j is on the right of vessel i in V, then vessel i *does not* "see" vessel j on its LEFT-DOWN view.

Figure 3 shows the LEFT-UP and LEFT-DOWN views of vessel j, with respect to a fixed berthing plan. It is clear that given any berthing plan, we can construct a pair (H, V) (need not be unique) satisfying the above properties. For any two vessels a and b, the ordering of a, b in H, V essentially determines the relative placement of vessels in the packing. For the rest of the paper, we write $a <_H b$ (and $a <_V b$) if a is placed on the left of b in H (resp. in V).

- If $a <_H b$ and $a <_V b$, then a does not see b in LEFT-DOWN or LEFT-UP, i.e., vessel b is placed to the right of vessel a in the berthing plan. In other words, vessel b can only be berthed after vessel a leaves the terminal.
- If $a <_H b$ and $b <_V a$, then a does not see b in LEFT-UP and b does not see a in the LEFT-DOWN view, i.e., vessel b is berthed below vessel a in the terminal.

For any H and V, either one of the above holds, i.e., either vessel a and vessel b do not overlap in time (one is to the right of the other) or do not overlap in space (one is on top of the other).

Note that every sequence pair (H, V) corresponds to a *class* of berthing plans satisfying the above properties. The constraints imposed by the sequence pairs splits into two classes: constraints of the type $x_i + l_i \leq x_j$ (in the space variables) or of the type $t_i + p_i \leq t_j$ (in the time variables). In this way, finding the optimal packing in this class, given a fixed sequence pair, decomposes into two subproblems: *space* and *time* (cf. Fig. 4). In the space and time graphs, each vessel is represented by a node and the constraints imposed by the sequence pair are represented by directed arcs.

Given a fixed sequence pair, it will be ideal if the optimal berthing plan can be obtained in a LEFT-DOWN fashion, i.e., berthing each vessel at the *earliest time* and *lowest possible position* available, subject to the sequence-pair condition. This method is advantageous since the berthing plan for vessels can actually be constructed greedily in an iterative manner, allowing us to handle a host of side constraints in real terminal berthing operations.

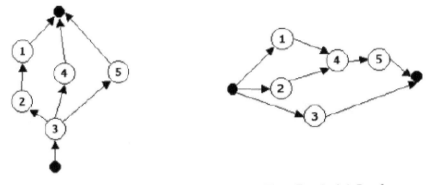

Space Constraint Graph   Time Constraint Graph

Figure 4. Directed graphs on the space and time variables arising from the sequence pair, H:{1,2,4,5,3} and V:{3,2,1,4,5}

Unfortunately, a LEFT-DOWN packing obtained with a fixed sequence pair may not always give rise to the optimal packing, due to the step-wise nature of the berthing space cost.

3.2   Time Cost Minimization with Fixed Sequence Pair

Given (H,V), let G_T be the directed graph associated with constraints involving the berthing time variables t_i. The time-cost problem can be formulated as:

$$\min_{t_i} \sum_{i=1}^{N} w_i \times f_i(t_i)$$

$$s.t. \quad t_i \geq r_i \ \forall \ i = 1,\ldots,N$$

$$t_i + p_i \leq t_j \ \text{if } (i,j) \in G_T.$$

Since $f_i(t_i) = (t_i - r_i - 2)^+$ is a non-decreasing function in t_i, the optimal solution to the above is clearly to set t_i as small as possible, subject to satisfying all the constraints. This is the dual version of the classical longest path problem in an acyclic digraph, and can be solved easily using a simple dynamic program. The t_i for the source node is set to the smallest possible value (i.e. r_i), and for a non-source node j, if S is the set of predecessors in G_T, then

$$t_j = \max\left(r_j, \max_{i \in S}(t_i + p_i)\right).$$

These values can be computed in a recursive manner (cf. Imahori et al. (2003) for an efficient dynamic programming based algorithm and Ahuja et al. (1993) for a general discussion on the longest path problem). Each t_i is selected to be the *smallest* possible value satisfying the constraints and the solution can be constructed in a recursive manner.

3.3   Space Cost Minimization with Fixed Sequence Pair

Given (H, V), let G_S be the directed graph associated with constraints involving the berthing location variables x_i. Let

$$L_l, U_l, l = 1, ..., W$$

denote the position of the lower and upper end of wharf l in the terminal. The space-cost problem can be formulated as:

$$\min \sum_{i=1}^{N} c_i(x_i)$$

$$s.t. \quad x_i + l_i \leq \sum_{l=1}^{W} U_l y_{il} \ \forall \ i = 1, \ldots, N$$

$$x_i \geq \sum_{l=1}^{W} L_l y_{il} \ \forall \ i = 1, \ldots, N$$

$$\sum_{l=1}^{W} y_{il} = 1 \ \forall \ i = 1, \ldots, N$$

$$x_i + l_i \leq x_j \ \text{if} \ (i, j) \in G_S$$

$$y_{il} \in \{0, 1\}, \quad l = 1, \ldots, W, \ i = 1, \ldots, N$$

$$x_i \geq 0, \quad i = 1, \ldots, N.$$

In the special case where $c_i(\cdot)$ is a constant function, the space cost minimization problem can be solved efficiently as in the previous case, using a similar Dynamic Program on G_S. In this instance, each vessel is berthed at the lowest possible position satisfying the precedence constraints and wharf-crossing constraints. For general $c_i(\cdot)$, unfortunately, we need to consider all possible berthing locations for vessel i in search of better berthing cost. This is the major bottleneck in the search for an optimal solution in this problem.

In fact, berthing the vessel according to the lowest possible position satisfying the constraints may not always give rise to good packing: i.e.,

LEFT-DOWN packing will not produce a good solution in terms of space cost.

To address this problem, we outline a method where the space cost minimization problem can be handled implicitly by extending the search space. We exploit the observation that the space cost function for each vessel is essentially a step function that depends on the number of berths in the terminal. Note that the number of berths in a terminal is relatively small (compared to the number of vessels). Furthermore, the space objective cost-function is constant within a berth.

We introduce a set of *virtual wharf marks* W, with $w \in W$ representing a vessel with $r_w = 0$, $l_w = 0$, $p_w = 0$, with additional constraint of the type for

$$x_w \geq \sum_{j=1}^{k(w)-1} b_j$$

some suitably selected berth $k(w)$. Note that b_j denotes the length of berth j. The constraint imposes a natural lower bound on the berthing location of the virtual wharf mark and indicates the decision to berth the virtual wharf mark at a location in the terminal not lower than berth $k(w)$.

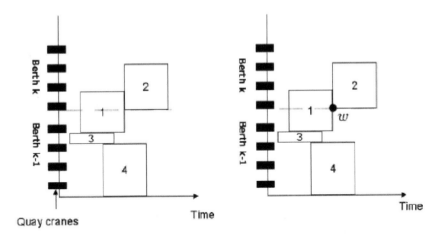

Figure 5. Berthing plan with virtual wharf mark as shown on the *right*. The sequence pair changes from (1,234, 4,312) to (12w34, 431w2)

Consider the berthing plan on the left side of Fig. 5. Due to the space cost function, it may be optimal to allocate vessel 2 to berth k. Unfortunately, a LEFT-DOWN packing will allocate vessel 2 to a location immediately above vessel 4, putting vessel 2 in berth $k - 1$. To rectify this situation, we introduce a virtual wharf mark and consider the modified sequence pair (12w34, 431w2). The left-down packing approach will give rise to the desired packing on the right of Fig. 5. The introduction of the virtual wharf marks in the sequence pair and the added constraints on the position of the wharf marks allows us to incorporate gaps into the berthing position of the vessels, to prop vessels up to certain berth. Note that the addition of the virtual wharf mark has no impact on the berthing time decisions.

It is not known, apriori, how many wharf marks will be needed to prop up the vessels to the desired locations in the terminal. But we note that for a sequence pair, with the right number of wharf marks distributed appropriately, we can obtain the optimal packing by using the LEFT-DOWN packing strategy. We record the observation in the following theorem.

Theorem 1. Suppose P^* is the optimal berthing plan to the problem, minimizing the total berthing time and space cost. Then there exists a set of virtual wharf marks $\{ w_1,...,w_K \}$, for 13 some K, such that P^* is equivalent to a berthing plan Q^*, involving all the vessels and virtual wharf marks, such that Q^* is obtained from a corresponding LEFT-DOWN packing using some sequence pair H^{**} and V^{**}.

In general, finding the optimal number of wharf marks and their associations with the vessels is extremely difficult. It is unnecessary to find the optimal packing at every stage because the sequence pair may not correspond to the optimal solution. The main advantage of this approach is that given *any* feasible packing, by introducing virtual wharf marks, we can encode the packing as one obtainable from LEFT-DOWN from an associated sequence pair in an enlarged space. This allows us to explore more complicated neighborhoods in the simulated annealing procedure.

3.4 Neighborhood Search Using Simulated Annealing

The approach outlined in the earlier section gives rise to an effective method to obtain a good packing, given a fixed sequence pair. We next use a simulated annealing algorithm to search through the space of all possible (H, V) sequence pairs.

The critical aspect for getting good solutions is to define an appropriate cooling schedule and a sufficiently large neighborhood that can be explored efficiently. In our empirical experiments, an exponential cooling schedule works well. With regard to the neighborhood, which is vital to ensure extensive exploration, we use the following structures:

(a) **Single Swap:** This is obtained by selecting two vessels and swapping them in the sequence by interchanging their position. Single swap is defined when the swap operation is performed in either H or V sequence.

(b) **Double Swap:** Double swap neighborhood is obtained by selecting two vessels and swapping them in both H and V sequences.

(c) **Single Shift:** This neighborhood is obtained by selecting two vessels and sliding one vessel along the sequence until the relative positions are changed; i.e., if $i, j, ..., k, l$ is a subsequence, a shift operation involving i and l could transform the subsequence to $j, ..., k, l, i$. There are many variants of this operation depending on whether vessel i (or l) is shifted to the left or right of vessel l (or i). We define single shift as a shift operation along one of the sequences.

(d) **Double Shift:** This defines the neighborhood obtained by shifting along both H and V sequences.

(e) **Greedy Neighborhood:** Given a sequence pair H and V and the associated packing P, we evaluate all possible locations that vessel i can take, with the rest of the vessels fixed in their respective positions. If there is a better location for the vessel, then we set the berth location of i to its new location. Note that since the time is kept constant, it is easy to check whether there is an overlap along the space dimension. Once the vessel is placed in the new location, we repeat the procedure for the rest of the vessels, until no improvement is possible.

Figure 6 shows examples of the above operations and their impact on the packing. The above neighborhoods described are simple perturbations of the sequence pair, but they result in remarkably different packing when compared visually. Due to the non-linear space cost, we observe that shifting of vessels between different berths may provide improvement in the solution. This motivated us to define the next class of neighborhood structure.

Figure 7 shows the packing and the corresponding sequence pair obtained from a simple greedy neighbor.

The greedy neighborhood artificially modifies the position of the vessel along space. Hence we may have cases where some vessels are placed on top of other vessels but none of the arcs in the space graph result in binding constraints. Fortunately, since the berthing space cost is constant within every berth, we have the following conditions for the berthing location for each vessel after the greedy modification: (1) the vessel i physically sit on top of another vessel, i.e., $x_i = x_j + l_j$ where j is a vessel that overlaps with i along time; or (2) $x_i = \sum_{l=1}^{k-1} b_l$ for some berth k.

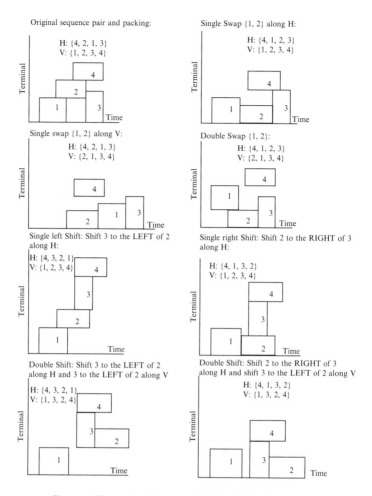

Figure 6. Examples of swap and shift neighborhoods

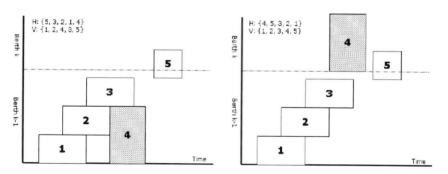

Figure 7. An example for greedy neighbor

In case (2) above, we introduce a virtual wharf-mark to the sequence pair to create the constraint that the vessel should be moored above berth edge k.

Implementation incorporating virtual wharf mark: The problem in adding virtual wharf marks as additional vessels in the search space is that it increases the problem size and hence the computation time. Here, we propose a cost-effective way of implementing the approach by employing dynamic lower bounds. Note that the vessels need to be propped after we employ the greedy neighborhood. Instead of adding virtual wharf marks, the idea is to (dynamically) set lower bounds for the berthing location for those vessels that need to be propped by a virtual wharf mark in the packing. We retain these lower bounds while exploring the neighborhood using operators (a)–(d), and change the lower bounds only when operator (e) changes the packing and introduces new virtual wharf marks. Searching the greedy neighborhood is computationally more expensive than simple sequence pair manipulation and hence, for the experiments, the operators (a)–(d) are employed successively while, the operator (e) is used whenever the other operators get stuck in a local optima.

The dynamic lower bounding technique described above is equivalent to adding a virtual wharf mark w to i (with w coming immediately after i in the H sequence, and w immediately before i in the V sequence), and performing all neighborhood searches treating iw in H, and wi in V, as "virtual" vessel. Note that swapping or shifting iw with j in H, or swapping or shifting wi with j in V, has the same effect of swapping or shifting i with j in the original H sequence, but maintaining a lower bound (determined by virtual wharf mark w) on vessel i in the neighborhood.

3.5   Rolling-Horizon Berth Allocation Planning

In the dynamic berth allocation planning model, we have the additional challenge of rolling the plan forward every time we move forward one period. This raises an important question: How do we handle the situation when certain vessels are already moored in the terminal? Note that the space allocated to these vessels cannot be used to moor other vessels.

We model these constraints as "forbidden zones" along the left edge of the time space network. In the presence of infeasible regions in the packing area, to obtain one-to-one correspondence between the sequence pair and the packing, we use a variant of the lower bounding strategy, which can be viewed as an extension of the virtual wharf mark approach. In the earlier results by Murata et al. (1998), infeasible regions were modeled as pre-placed vessels and a greedy strategy was used to modify a packing to ensure that pre-placed vessels are restored to their original locations. But with a complexity of $O(N^4)$ for instances with N vessels, it is computationally expensive. Herein we develop a strategy to specifically cater to the issue of packing during dynamic deployment. Fig. 8 shows a typical scenario.

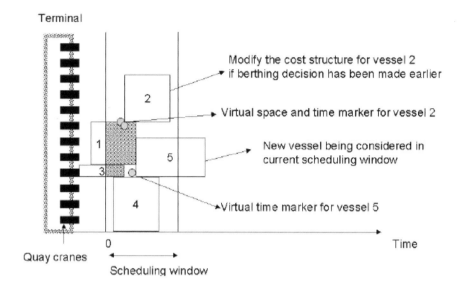

Figure 8. Handling pre-placed vessels

Along space, if a vessel does not have binding arcs in the space graph G_S, then we set an appropriate lower bound for the berthing location decision of this vessel. Along time, if the vessel is adjacent to an infeasible region and there are no binding arcs in the time-graph, we can also set an appropriate lower bound to the berthing time decision of these vessels.

An initial packing is obtained using a greedy insertion strategy. The neighborhoods are then explored using the previous neighborhood structures, with the additional lower bound constraints on berthing time and location decisions. We omit the details of the implementation here.

Revisions to berthing plan: The ability to build-in the latest arrival information and to revise plans is definitely an advantage to the terminal operator in a dynamic setting; but frequent revision of the berthing plan is not desirable from a resource planning perspective. This is due to the fact that, based on the berthing plan, the port will normally have to plan the yard load and activate needed resources within the terminal to support loading/discharging activities from the vessels. In a container port, the following major resource-hogging events are affected while servicing a particular vessel: Quay Cranes, Yard Cranes, Prime Movers, Operator-Crew, Container Movement, etc. Typically, the container and resource management in the yard is a complex problem and is done beforehand based on the berth plan. Hence, any major deviation from the plan in a vessel's location or service time, results in a major re-shuffle of crew and prime mover allocations. It also results in re-deployment of cranes. Further, last minute changes to the berthing plan and re-deployment of personnel and equipment lead to confusion. Hence it is preferable that a proposed berth plan can be executed with minimal changes.

To ensure *steadiness* (i.e., minimize frequent revisions) in the berth plan, the terminal operator can opt to freeze the berthing location decision made during earlier scheduling windows. Another alternative is to penalize deviation from an earlier plan by imposing a large penalty for changing the berthing location or time for a vessel. We opt for the second strategy, where the space and time cost function for vessel i is adjusted to

$$c_i'(x) = \begin{cases} 0 & \text{if } x = x_i, \\ A & \text{otherwise}, \end{cases}$$

$$f_i'(t) = A(t - t_i)^+,$$

where x_i, t_i are the earlier berthing decision, and A is a huge penalty term. The choice of $c_i(\cdot)$ and $f_i(\cdot)$ essentially forces vessel i to be moored at the original berthing location and time.

4. Throughput Consideration

The previous sections outlined a method to allocate berthing space to vessels in a rolling horizon framework. However, it does not preclude the possibility that certain classes of vessels will be delayed for an excessive period, and is thus lost from the system. This indirectly reduces the berth utilization and throughput of the terminal, which is a primary concern in berth planning. In this section, we propose a strategy to redistribute load within the terminal at periodic intervals, to ensure that the throughput of the terminal will be maximized. While the concerns with throughput can also be addressed using a more complicated cost function $f_i(t)$ to penalize excessive delays, finding the appropriate choice of the penalty function proves to be tricky in practice, in view of the fact that there are different priority classes of vessels waiting to be berthed at the terminal. The penalty cost for inducing additional delay should depend also on the priority classes of the vessels competing for space at the same time. Our approach here is neater and builds on the insights of the structure of provably throughput optimal policies in stochastic processing network.

To study the issue of throughput in the berth allocation planning problem, we first discretize the berth space into integer units, indexed by $k = 1,2,...,K$. Each unit of the space is called a section. Each section can accommodate one vessel each time, and each vessel occupies an integer number of consecutive sections when it is moored. For ease of exposition, we group the vessels into categories or *classes* by the number of sections they require.[22] All the vessels of the same class require the same number of sections. The classes of vessels are indexed by $i = 1,2,...K$. We denote $Z_i(t)$ to be the number of class i vessels that have arrived at the terminal either waiting to be served or being served at time t. $Z(t) = (Z_i(t), i = 1,2,...,I)$ is called the buffer level at time t. We assume the vessels in the same class are moored in a FCFS fashion. Therefore the berth planning problem is to decide how to assign the sections to classes. We call each possible assignment of the sections to classes as an *allocation* denoted by a. We denote the number of class i vessels moored by allocation a as $n_i(a)$.

4.1   Discrete Maximum Pressure Policies

In this section, we propose a family of policies called *discrete maximum pressure* policies for berth allocation given the processing times are bounded by some constant U. Define the pressure of a berth allocation network with buffer level z under allocation a to be

$$p(a, z) = \sum_i n_i(a)\mu_i z_i,$$

where $\mu_i = 1/m_i$ and m_i is the average length of port stay for class i vessels.

The discrete maximum pressure policy for berth allocation with length $L \geq U$ is defined as follows.

Discrete Maximum Pressure Policy for Berth Allocation

- The time horizon is divided into cycles of length L.
- At the beginning of each cycle, an allocation with *maximum pressure* (i.e., $\arg\max_a p(a, Z(t))$) is selected, and each section is assigned to the vessel class given by the allocation until

   1. There is no vessel of the corresponding class or
   2. It finishes a job and can not finish the next one within the cycle.

---

[22] In practice, these vessels may differ in preferred berth allocation. However, as we will focus on the throughput attained in this section, we do not distinguish between these vessels.

In both cases, the section is released and is ready to be assigned to any vessel that can be finished within the cycle.

- One has flexibility to assign the released sections. We can allocate space from the released section according to the objective outlined in the previous section, or we can do so such that the number of the remaining idle sections is minimized.

Note that if a vessel arrives near the start of any planning cycle, it may not be served even if there is space available in the terminal. This happens when its port stay time is long and hence it will not depart within the current planning cycle. In this case, instead of berthing the vessel, the space is reserved till the next planning cycle for redistribution of load, to accommodate the allocation with the maximum pressure. In a way, the discrete maximum pressure policy is similar to the intuitive approach of clearing the buffers whenever there are many vessels waiting in the system. It uses the notion of $p(a, z)$ to operationalize this concept.

Discrete maximum pressure policies are a variant of the maximum pressure policies (cf. Dai and Lin (2005)). The maximum pressure policies were proven to be throughput optimal for a class of stochastic processing networks. That is, the load can be handled by maximum pressure policies if it can be handled by any other scheduling policy. One can model the berth allocation problem as a stochastic processing network and prove similarly that, by appropriately choosing the length of the planning cycle, the discrete maximum pressure policies can handle the load that can be handled by any other scheduling policy. We omit the detailed proofs in this paper due to space limitation. The argument can be adapted from the proof of throughput optimality of maximum pressure policies given in Dai and Lin (2005). In the following, we point out the difference between the discrete maximum pressure policies and the maximum pressure policies.

- For maximum pressure policies, the allocation is updated every time when there is an arrival or a service completion, while for discrete maximum pressure policies, the allocation is updated only when all vessels are served from some class or the planning cycle is over.
- Maximum pressure policies are in general a preemption allowed policies, but service is not to be interrupted for discrete maximum pressure policies.

4.2   A Numerical Example

To illustrate how the discrete maximum pressure policy works, consider the application of this policy to the problem described in Sect. 1. We assume $U = 16$, and set $L = 160$. Figure 9 compares the buffer build up in the system when the discrete maximum pressure policy is employed against a naive static berth planning system in a rolling horizon framework. Firstly, it is observed that under the discrete maximum pressure policy, the system is stable with the dynamics repeating every 1,920 h as opposed to the static berth allocation system where the buffer builds up indefinitely. This shows the importance of the dynamic policy in a congested scenario. Secondly, at the beginning of every cycle (i.e. with length L), the system needs to be cleared. This purposely induces delays in the vessels and also increases the size of the static berth planning problem to be solved. The spike in the number of vessels waiting to be serviced at the beginning of each cycle as shown in Fig. 9 and the increase in delay experienced by the class 1 vessels (Two vessels are delayed by 5 and 1 h respectively every 160 h, up from 0 h when employing the static allocation alone) support this observation.

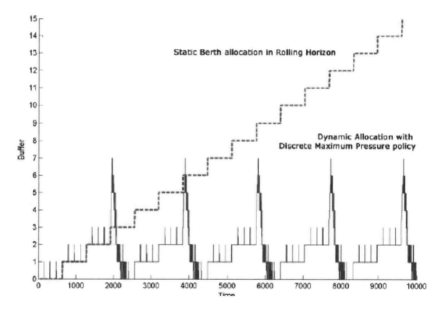

Figure 9. In a heavy load setting, the discrete maximum pressure policy ensures that the system remains stable. However, redistributing the load at the beginning of every cycle increases the number of vessels waiting to be serviced at that time

In a moderate traffic scenario with a large number of classes of vessels, the buffer build up at the beginning of each cycle may potentially be extremely large, making the static berth planning problem computationally expensive. Further, under moderate load, we would hope that the issues of infinite buffer build up to be eliminated by employing a good static berth planning system and looking further ahead of the scheduling window. Hence, it would be advisable to select the load at a terminal such that no vessel will be substantially delayed. It would be recommended to employ the discrete maximum pressure policy only in situations where heavy load scenarios cannot be avoided.

5. Computational and Simulation Results

5.1 Computational Results: Static Berth Allocation Planning Model

To evaluate the performance of the proposed approach, we need to compare the performance against a suitable lower bound. To this end, we use tools from mathematical programming to obtain a lower bound to our berth planning problem.

Let C_i be the set of allowed positions that vessel i can be moored in the schedule. The set C_i takes care of all the constraints in berthing positions (e.g., no mooring across wharfs) and berthing times (e.g., not earlier than arrival) of vessel i. We can immediately construct a lower bound to the problem by solving the following:

$$\max_{\pi(x,t)} \min_{(x_i,t_i)\in C_i} \sum_{i=1}^{N}\left(c_i(x_i)+\omega_i f_i(t_i)+\int_{x_i}^{x_i+l_i}\int_{t_i}^{t_i+p_i}\pi(x,t)dxdt\right)-\int_{0}^{B}\int_{0}^{\infty}\pi(x,t)dxdt.$$

The variables $\pi(x,t)$ are the dual prices associated with the non-overlapping constraints (for the position (x,t) in space). With fixed $\pi(x,t)$, the inner optimization problem is trivial. We use a version of subgradient algorithm, known as the volume algorithm in the literature (cf. Barahona and Anbil (2000)), to update the dual prices and to solve the above problem. To make the problem manageable so that the lower bound can be obtained within reasonable time, the space–time network is also discretized to moderate sizes. Note that the discretized version of the problem will still provide us with a lower bound for the static berth planning problem.

In this section, we compare the performance of the virtual wharf mark approach with that of the lower bound. We also show the difference in computational time taken to solve the problem. We report computational results on problem sizes including 30, 50, and 70 vessels. The instances have varying Resource Utilization (RU) levels. Average RU is a measure of traffic load at the terminal and the max RU measures the peak demand for berthing space within the scheduling window. For each problem size, we compare the computational performances under two different scenarios:

- Congested Scenario: In this situation, all vessels available for packing arrive at time zero, thus creating intense competition for usage of the terminal. The instances in this scenario are created with high average RU.
- Light Load Scenario: In this case, all vessels can be moored immediately upon arrival and at the most preferred berth. We do this working backward, i.e., starting with a packing, we devise cost parameters to the problem so that the packing is indeed the optimal solution. This is constructed by choosing appropriate values for vessel arrival time and preferred berthing locations. The average RU in these instances is typically lower and the max RU is always less than 1.

The computational performance for the experiments (averaged over ten random instances) is summarized in Tables 1 and 2. Note that the proposed lower bound is understandably weak, since the Lagrangian relaxation model relaxes the non-overlapping constraints in the problem. The simulated annealing (SA) algorithm is still able to return a solution close to 25–36% of this lower bound (LB), depending on whether space cost is included in the model or not. However, we believe that the wide gap between SA and LB is primarily due to the weak lower bound. This is confirmed by the light load cases, where the simulated annealing algorithm is able to consistently return close to optimal solution in all ten random instances (0.33% vessels delayed and <6.3% vessels placed in inferior locations). Note that in the light load scenario, we know the optimal solution (0 vessels delayed, 0 vessels placed in inferior locations).

Table 1. Comparison with lower bound: congested scenario

# Vesse ls	Avg RU %	Max RU %	Space cost	Lower bound (LB)	Running time (s)	Heuristic solution (sA)	Runnin g time (s)	(SA-LB)/LB
30	60	250	No	613.8	52	794.3	15	0.32
			Yes	1,645.6	54	2,186.4	16	0.36
50	80	420	No	2,119.6	69	2,769.1	148	0.31
			Yes	2,645.4	69	3,468.9	155	0.33
70	75	580	No	5,588.6	165	7,260.3	643	0.31
			Yes	6,503.1	165	8,078.8	661	0.25

Table 2. Performance of simulated annealing: light load scenario

# Vessels	Avg RU %	Max RU %	Vessels delayed (%)	Vessels in inferior berth (%)
30	25	60	0.33	2.67
50	28	65	0.00	3.40
70	37	82	0.29	6.29

5.2 Simulation Results: Rolling Horizon Berth Allocation Planning

There are many variables that affect the dynamic deployment scenario and the impact can only be studied by using a simulation model.

Arrival pattern: We generate arrival to the terminal using an arrival pattern representative of a typical port (cf. Fig. 10). There are 48 vessels scheduled to call at the terminal on a weekly basis and the figure shows the expected time of call during a week for each vessel. There are several other vessels shown below the chart, and these correspond to vessels which do not call weekly at the terminal, but may call every 10 days.

Design of experiment: The following describes the events regarding vessel arrival times and the subsequent planning and execution that is done prior to berthing a vessel. The experimental setup for the simulations also follows these events.

- The vessels follow a cyclic arrival with a period of 7 (regular calls) or 10 (irregular calls) days.
- The arrival time according to the template is the *Initial Berth Time Requested* (IBTR) by the vessels. The actual arrival time is stochastically distributed with the IBTR as the mean arrival time.
- The vessels are randomly partitioned into seven priority classes, and the preferred berth location is generated according to the berthing location within the template in Fig. 10.
- Typically, to ensure *steadiness*, the berth locations are fixed (or minimum changes made) for the vessels set to arrive in the nearest shifts. This is done to prevent last minute resource re-allocation, which may lead to inefficient resource management. Steadiness is measured based on changes in the plan and the execution within the next two shifts (16 h) from the current time. Due to uncertainty in arrival time, the measure is based only on changes in vessel location. In the simulation, to ensure 100% steadiness, we will also impose upper and lower bound constraints on the berthing location of the vessels.

The data template as shown in Fig. 10 is clearly not heavily loaded since there is enough time in-between vessel arrivals to prevent delay propagation. Hence, in the computational experiments A–F, we avoid the discrete maximum pressure policy and the associated re-balancing exercise and analyze the dynamic performance of the proposed berth planning system. In Experiment G, we present a congested scenario and show the impact of the discrete maximum pressure policy on the throughput of the system.

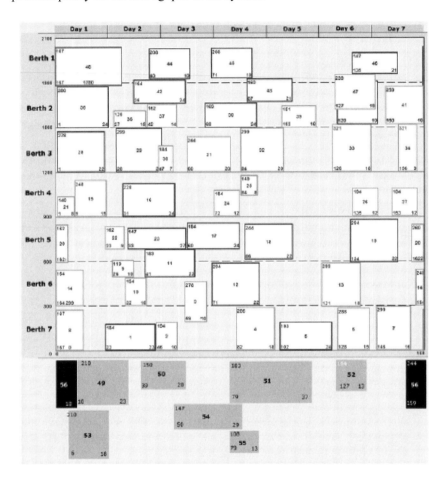

Figure 10. Template for the simulation

We specifically run the following test scenarios and compare the performance of the dynamic berth allocation planning model in a rolling horizon setting. For comparison, we generate the IBTR and the UBTR beforehand and run the following experiments. There are some managerial issues that are interesting to study in the berth planning problem. The developed simulation package allows us to study the impact of these on the overall performance of the system. For instance, we test the effect of the *planning time window*, i.e., how far to look ahead in the scheduling window at each period. There is a trade-off between myopic planning (small time window) and inaccuracy in data (long time window). We also test the trade-off between the objective of attaining high BOA for vessels, and the desire to minimize re-planning of activities.

Experiment A: Forecast is accurate. In this experiment, the actual arrival time of vessels follows the data as given, i.e., the IBTR corresponds to the actual arrival time. This test represents the current arrival scenario and provides a bench mark for the performance of the proposed system. Ideally, if we use the 48 vessels scenario in Fig. 10, and if the arrival time is accurately reflected, we expect all vessels to attain BOA status and to be berthed at their preferred location. Table 3 shows the results from a run of the simulation package, with a planning time window of 48 h. Furthermore, berthing plan for vessels due to arrive over the next 16 h (where arrival time is known exactly) is frozen.

Table 3. Simulation results obtained for Experiment A

Class	# of vessels served	% BOA	Average delays (h)	% Allocated preferred berth
0	126	90	0.38	79
1	56	75	0.50	50
2	56	100	0.00	100
3	84	100	0.00	82
4	70	100	0.00	97
5	210	100	0.00	87
6	56	100	0.00	96
7	14	100	0.00	93

Ninety-six percent out of 672 vessels served in the 14-week simulation environment received BOA treatment, and close to 85% were moored at their designated preferred berths.

The above results indicate that the proposed dynamic berth allocation planning model performs extremely well in this environment.

Experiment B: Template-based data with uncertainty in arrival time. In this setting, we assume that the actual arrival time has variance 30 and 10 with respect to IBTR and UBTR respectively. We examine the impact of the planning window on the performance of the algorithm. Again, the berthing plan for vessels due to arrive over the next 16 h (where arrival time is known exactly) in each planning time window will be frozen. We summarize the simulation results in Table 4.

Table 4. Simulation results obtained for Experiment B

Windows	% BOA	Average delays (h)	% Allocated preferred berth
16	88	0.81	70
24	89	0.53	75
48	90	0.72	78
72	93	0.33	76

Interestingly, the results show the advantage of incorporating the longer planning time window of 72 h over the shorter planning time window of 16 h. Advance information on vessel arrival time, despite the uncertainty in the information, is still beneficial for the terminal in terms of berth planning. This is mainly because the plan for the first 16 h will be frozen, and hence advance arrival information beyond the first 16 h helps the terminal to obtain better berthing plan.

Experiment C: Effect of freezing berthing plan. In this setting, we compare the performance of the model under the same setting as experiment B, except that the port will re-plan and possibly change berthing positions every time there is an update on vessel arrival data, i.e., we do not freeze the berthing plan in every planning window.

Table 5. Simulation results obtained for Experiment C

Windows	% BOA	Average delay (h)	% Allocated preferred berth	% Steadiness
16	94	0.25	78	23
24	94	0.23	75	29
48	94	0.26	73	31
72	93	0.29	75	38

In this case, the ability to re-plan allows the model to achieve better performance in terms of BOA and preferred berth allocation statistics. Furthermore, the advantage of a longer planning time window is no longer apparent. By planning using only the accurate arrival information (within first 16 h), and with the capability to update the berthing plan every hour, the terminal can achieve an equivalent performance to the case when longer planning time windows are used. However, this comes at a price of frequent berthing plan revisions, which is undesirable. In Table 5, steadiness is a measure of the number of vessels that remain at the same location within the 16-h period. In Experiment B, steadiness is 100%.

Experiment D: Effect of differentiated priorities. In this experiment, we study the effect of differentiated priorities on the performance of the system, using a planning time window of 48 h. We compare average BOA and preferred berth allocation between the current system (with differentiated priorities), and the case when all vessels are accorded high or low priority status. Table 6 summarizes the results observed.

Table 6. Simulation results obtained for Experiment D

Priority	% BOA	Average delay (h)	% Allocated preferred berth	% Steadiness
Differentiated	94	0.26	73	31
Equal-high	97	0.14	67	30
Equal-low	90	0.58	86	46

Note that in the equal-high scenario, the overriding concern is to berth all vessels on arrival. In the equal-low scenario, the focus is on maximizing the benefits of preferred berth allocation. As in experiment C, we do not freeze the berthing plan in this case. The results indicate a good balance between the desires to obtain high BOA with a high percentage allocated to preferred berth.

Experiment E: Irregular arrivals. Here, we repeat the experiment set up in A, but with a loading profile according to Fig. 10, including the vessels with a cycle time of 10 days. Table 7 shows the results obtained with the additional classes of vessels.

Table 7. Simulation results obtained for Experiment E

Class	# of vessels served	% BOA	Average delays (h)	% Allocated preferred berth
0	126	86	2.2	83
1	56	61	1.57	52
2	56	77	1.88	88
3	104	98	0.13	86
4	100	98	0.04	88
5	220	98	0.04	82
6	56	100	0	91
7	34	100	0	68

Note that in this case, berth utilization of the terminal increases at a price: while class 6 and 7 vessels continue to enjoy BOA berthing, the service experienced by class 0–5 vessels suffers. This is especially apparent for class 0–2 vessels, where the service standard drops by close to 20%.

Experiment F: Irregular arrivals and uncertainty in arrival time. Here, we repeat the experiment set up in B, but with an irregular vessel arrival pattern along with uncertainty in arrival time. Table 8 shows the results obtained, which conform to the results in experiment B.

Table 8. Simulation results obtained for Experiment F

Windows	% BOA	Average delays (h)	% Allocated preferred berth
16	83	1.47	73
24	86	0.91	80
48	88	0.78	77
72	86	0.93	75

Experiment G: Scheduling under congestion. In this experiment, we study the impact of the scheduling policy on the throughput of the system under a congested scenario. Congestion in a port is typically managed so that no class of vessels suffer throughput loss. However, when certain berths or wharfs are under maintenance or upgrade, the usable length of the terminal gets restricted which results in increased congestion when the arrival pattern remains unaltered.

We use this scenario to create a realistic data set by modifying the terminal length for the arrival pattern in Fig. 10. To measure throughput loss, we restrict the maximum buffer for each vessel class to 4 and the vessels arriving after the buffer is full are categorized as *lost sales.*[23]

Simulations were performed by varying terminal length settings (from the existing 2,100 m long terminal to 1,600–1,700 m) and we compare the lost sales, throughput and delays when scheduling with the naive rolling horizon setup as opposed to employing the discrete maximum pressure policy. The simulation was performed for 3,300 h generating 828 vessel arrivals and the load was redistributed using the discrete maximum pressure policy every 400 h.

The progression of lost sales with time is shown for one of the simulations (with terminal length of 1,600 m) in Fig. 11 and similar progression was observed for the remaining simulations. On an average over three simulations, employing the discrete maximum pressure policy reduced the lost sales by ten vessels, a near 30% improvement over a naive rolling horizon framework.

The average throughput in the system increased from 95 to 97% while using the discrete maximum pressure policy as shown in Fig. 12. However, the redistribution procedure results in increasing the average delay experienced by the vessels from 49.3 to 86.3 h. Note that * in the Fig. 12 indicates that a particular class of vessel was never served in one of the simulations employing the naive rolling horizon framework, and the average delay excludes this class.

---

[23] Computing the actual throughput in a congested scenario proves to be tricky, as the BAP algorithm becomes very slow when there is huge number of vessels that need to be scheduled at each planning epoch. The lost sales provide a surrogate measurement for throughput, and prevent an excessive number of vessels building up at each planning epoch.

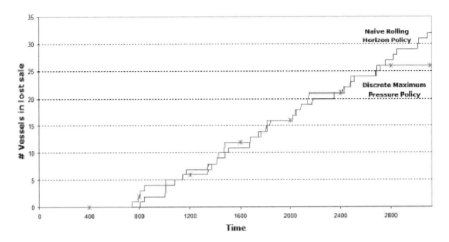

Figure 11. Plot comparing the lost sales while employing the naive rolling horizon policy against a discrete maximum pressure policy

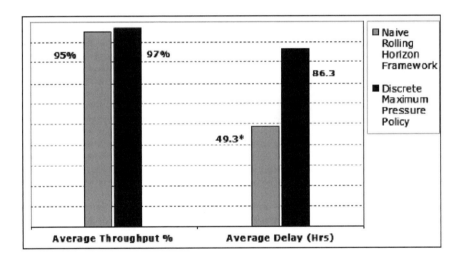

Figure 12. A comparison of average delay and throughput of the berth planning system

Although the redistribution procedure results in increased average delay, the improvement in throughput when employing the discrete maximum pressure policy is evident from the simulation results.

6.   Conclusion

In this paper, we proposed an efficient heuristic to construct a berthing plan for vessels in a dynamic berth allocation planning problem. Our method builds on a well-known encoding scheme used for VLSI design and rectangle packing problems. We show that the same approach can be used effectively for the berth allocation planning problem, with the introduction of virtual wharf marks. Extensive simulation results based on a realistic set of vessel arrival data shows the promise of this approach. For a moderate load scenario, this approach is able to allocate space to over 90% of vessels upon arrival, with more than 80% of them being assigned to the preferred berthing location. Our simulation results also show that the performance varies according to the various policy parameters adopted by the terminal operator. For instance, frequent replanning and revisions to the berthing plan, with the resulting undesirable impact on terminal resource deployment, yields close to 1–6% improvement in BOA statistics.

For a heavy load scenario, we need to ensure that the throughput of the terminal will not be adversely affected by the design of the berthing algorithm. We use the discrete maximum pressure policy to redistribute load at every L interval, where L is suitably selected based on the loading profiles at the terminal. Simulation results show that this strategy helps in the heavy traffic scenario to minimize throughput loss due to excessive waiting for certain classes of vessels.

There are, however, several limitations to our approach. Future work will try to address some of these limitations. For instance, we have ignored several important berthing constraints while allocating space to vessels. Some of the constraints ignored include the availability and scheduling of cranes and gaps between berthing vessels. We need to address these concerns in the static berth allocation planning problem so that the model can be applied in practice. Furthermore, we have also assumed historical crane intensity rates while working out the port stay time. In practice, the port stay time of each vessel can be erratic. We need to enrich the berthing algorithm to handle situations with uncertain port stay times.

Acknowledgments

The authors would like to thank Research supported in part by National Science Foundation grant DMI-0300599, National University of Singapore Academic Research Grant R-314-000-030-112 and R-314-000-048-112, TLI-AP, a partnership between National University of Singapore and Georgia

Institute of Technology, and by the Singapore-MIT Alliance Program in High Performance Computation for Engineered Systems.

References

R.K. Ahuja, T.L. Magnanti, and J.B. Orlin (1993). *Network Flows: Theory, Algorithms, and Applications*. Prentice-Hall, Englewood Cliffs, NJ.

F. Barahona and R. Anbil (2000). The volume algorithm: producing primal solutions using a subgradient method, *Mathematical Programming*, 87, 385–399.

G.G. Brown, S. Lawphongpanich, and K.P. Thurman (1994). Optimizing ship berthing, *Naval Research Logistics*, 41, 1–15.

C.Y. Chen and T.W. Hsieh (1999). A time–space network model for the berth allocation problem, Presented at the *19th IFIP TC7 Conference on System Modelling and Optimization*.

L.W. Chen (1998). Optimisation problem in a container port, M.Sc. Research Report, SoC, NUS.

J.T. Chia, H.C. Lau, and A. Lim (1999). Ant colony optimization for the ship berthing problem, in P.S. Thiagarajan, R. Yap (Eds.): *ASIAN'99*, LNCS 1742, pp. 359–370.

J.G. Dai and W. Lin (2005). Maximum pressure policies in stochastic processing networks, *Operations Research*, 53, 197–218.

A.V. Fishkin, K. Jansen and L. Porkolab (2001). On minimizing average weighted completion time: A PTAS for scheduling general multiprocessor tasks, in *Proceedings 13th International Symposium on Fundamentals of Computation Theory (FCT'01)*, Rusins Freivalds (Ed.), Riga, LNCS 2138, Springer Verlag, pp. 495–507.

Y.P. Guan and R.K. Cheung (2004). The berth allocation problem: models and solution methods, *OR Spectrum*, 26, 75–92.

Y.P. Guan, W.Q. Xiao, R.K. Cheung, and C.-L. Li (2002). A multiprocessor task scheduling model for berth allocation: heuristic and worst case analysis, *Operations Research Letters*, 30, 343–350.

S. Imahori, M. Yagiura, and T. Ibaraki (2003). Local search algorithms for the rectangle packing problem with general spatial costs, *Mathematical Programming Series B*, 97, 543–569.

A. Imai, E. Nishimura, and S. Papadimitriou (2001). The dynamic berth allocation problem for a container port, *Transportation Research Part B*, 35, 401–417.

A. Imai, E. Nishimura, and S. Papadimitriou (2003). Berth allocation with service priority, *Transportation Research Part B*, 37, 437–457.

K.H. Kim and K.C. Moon (2003). Berth scheduling by simulated annealing, *Transportation Research Part B*, 37, 541–560.

C.L. Li, X. Cai, and C.Y. Lee (1998). Scheduling with multiple-job-on-one-processor pattern, *IIE Transactions*, 30, 433–446.

A. Lim (1998). On the ship berthing problem, *Operations Research Letters*, 22(2–3), 105–110.

S.N. Loh (1996), *The Quadratic Assignment Problem and its Generalization to the Berth Allocation Problem*, Honours Years Project Report, DISCS, NUS.

K.C. Moon (2000). *A Mathematical Model and a Heuristic Algorithm for Berth Planning*, Unpublished Thesis. Pusan National University.

H. Murata, K. Fujiyoshi, S. Nakatake, and Y. Kajitani (1996). VLSI module placement based on rectangle packing by the sequence pair, *IEEE Transactions on Computer Aided Design of Integrated Circuits and Systems*, 15–12, 1518–1524.

H. Murata, K. Fujiyoshi, and M. Kaneko (1998). VLSI/PCB placement with obstacles based on sequence pair, *IEEE Transactions on Computer Aided Design of Integrated Circuits and Systems*, 17, 60–68.

Y.M. Park and K.H. Kim (2003). A scheduling method for Berth and Quay cranes, *OR Spectrum*, 25, 1–23.

D. Steenken, S. Voss, and R. Stahlbock (2004). Container terminal operation and operations research a classification and literature review, *OR Spectrum*, 26, 3–49.

X. Tang, R. Tian, and D.F. Wong (2000). Fast evaluation of sequence pair in block placement by longest common subsequence computation, in *Proceedings of the Conference on Design, Automation and Test in Europe*, pp. 106–111.

C.J. Tong, H.C. Lau, A. Lim (1999). Ant Colony Optimization for the ship Berthing problem, *Asian Computing Science Conference (ASIAN)*, pp. 359–370.

Merchandise Planning Models for Fashion Retailing

Kumar Rajaram

Decisions, Operations and Technology Management, The Anderson School at UCLA, 110 Westwood Plaza, Los Angeles, CA 90095-1481, USA

Abstract: Merchandise planning is the process conducted by a retailer to ensure that the right product is available to the customer at the right place, time, quantity and price. This process involves selecting the products the retailer will carry and determining the purchase quantities of these products. Merchandising has become more complex because of changes in the retail industry such as consolidation, global sourcing, higher levels of competition, increasing product variety, reduced life cycles and less predictable demand. Enhancements in information, manufacturing and distribution technology offer potential to reduce the large markdowns due to excessive inventory and lost sales opportunity due to sellouts currently prevalent in this industry.

In this chapter, we study two problems in retail merchandising. In the first problem, we develop a methodology to improve the accuracy of merchandise testing by choosing how many and at which stores to test new products and how to extrapolate test sales into a forecast for the entire season across the chain. In the second problem, we consider replenishment based on actual sales, a strategy that can be employed by the retailer to minimize inventory risk associated with an assortment of products. In both these problems, we develop models that are compared to existing practices at these retailers using real sales data. Comparing our techniques to current practices, we found they could reduce markdowns due to excessive inventory and lost margins due to stockouts enough to increase profits by over 100% in each application. General insights on improving this process and future research directions are described.

1. Introduction

Much has been written about the problem of high inventory costs due to supply-demand mismatches for products with extremely short life cycles and highly unpredictable demand. While this problem arises in a variety of industries, the apparel retailing industry provides a particularly fascinating context to study this problem due to several reasons. It is a high impact industry, a large proportion of whose products possess these characteristics and sellouts and overstock of merchandise due to these mismatches have a huge impact on operating costs. In 1986, these costs have been estimated by Frazier (1986) to be around $25 billion annually for the U.S. apparel industry, which represented around 25% of total retail sales. To better appreciate these costs, it is important to note that traditionally profit margins in this industry have been slender. For instance, the Standard and Poors industry surveys reports an average annual profit margins of 12.1% during the years 1991–1996 across four major retailers in the specialty apparel sector. The corresponding average annual Return On Assets (ROA) was 10.9%. Consequently, even small reduction in these costs could have a significant impact on improving profit margins and ROA.

Enhancements in production, distribution and information technology offer the opportunity to reduce the cost of these mismatches. Flexible manufacturing technology permits production of smaller batches of garments at a large variety enabling more accurate matching of customer preferences. Hammond (1992) describes a retailer that employs such techniques. Faster and more cost efficient distribution technologies have been developed and are being extensively utilized. Several retailers including Laura Ashley, an U.K. based women's specialty apparel retailer (Anthony and Loveman, 1994) and the American mega retailer WalMart (Stalk et al., 1992) have successfully implemented such technologies. Information technology is being used to understand customer preferences and integrate manufacturing with distribution technology to react rapidly to these preferences. Stalk et al. (1992) describes the integration of these technologies at Walmart. Surveys by the Economist (1995) and the Standards and Poors (1998) indicate that apparel retailers have invested heavily in these technologies. However, in the final analysis, the benefits of these technologies depend heavily on the process by which supply and demand is matched at a particular retailer.

The process used to match supply with demand in the retailing industry is commonly referred to as the "Merchandise Planning Process." Broadly speaking, the objective of this process is to ensure that the product is available to the customer at the right quantity, price, time and place (the so called four holy rights of retailing!). Major decisions that arise in this process include the choice of products, determining their order quantities, deciding how to source these products and distribute them to the stores and finally setting and if necessary changing their prices. Retailers have recognized the strategic importance of this process (Aufreiter et al., 1993). They have begun to focus on improving some of the decisions made in this process including deciding the assortment or which products to include (Fox, 1995) and ensuring that these products are available to the customer at the right quantity at the stores (Wilson et al., 1995).

Since retailers have always been making these decisions, to understand why they deserve additional and intense scrutiny, it is useful to look at the types of products offered by these retailers. To make these decisions across a wide range of products, retailers broadly classify their merchandise as basic and fashion. Basic merchandise like jeans, underwear, men's dress shirts, typically have more stable, predictable demand and longer life cycles than fashion products like women's dresses, jewelry, sportswear etc. which are typified by highly unpredictable demand and extremely short life cycles. This stability of demand and longer duration of the sales season for basics has enabled retailers to use advances in retail production, distribution and information technology in conjunction with conventional techniques of forecasting (like exponential smoothing) and inventory control to efficiently match supply with demand. It is highly unlikely that we would go to a store and not find a white shirt or find that shirt selling at a steep discount. This initiative in the apparel industry is commonly known as quick response. Hammond (1990) provides an excellent overview of quick response in the apparel retailing industry.

While the planning of basics has been well served by these techniques, they seem inadequate in dealing with the merchandising challenges posed by fashion products. The popular evidence for this assertion is abundant as reported by several newspapers (San Juan Star, 1996; USA Today, 1996) and that emphasize the high level of sellouts of some fashion products, while a majority of them are sold at a steep discounts across a variety of retailers. It is not uncommon to walk into a store and find that popular products are those which are not in stock and to find the same store cluttered with unpopular products selling at steep discounts.

To understand the impact of stockouts and markdowns at a particular firm, we analyzed data for a line of 300 dresses at a major catalog retailer over a 3-year period. When a dress is sold at a price below procurement cost, we incur overstock costs corresponding to the difference between procurement cost and the markdown price for that dress. Stockouts occur when a particular dress is not in stock and the sale is lost, an event that could be recorded due to the nature of catalog retailing. The corresponding understock cost per unit for this dress is the difference between full price and procurement cost. Table 1 summarizes the total number of dresses purchased at the beginning of the season, the total stockouts and overstock units in the season and the total understock and overstock costs (as % of sales). This suggests that total overstock and stockout costs across these 3 years are around 16% of sales, a large number when one recognizes that net profit margins at this retailer were only around 3% of sales. This table also provides us with some insight on how these costs are generated. In the first year, these costs are largely due to overbuying and the resulting overstock cost. In response, the buy is scaled down in the next year and now most of the costs are due to stockouts. This is corrected in the third year with a larger buy that once again results in a larger proportion of overstock costs.

Table 1. Merchandising performance at a catalog retailer

	1993	1994	1995
Total buy (units)	68,507	58,912	81,995
Stockouts (units)	912	29,005	4,457
Overstock (units)	29,417	25,532	33,120
Stockout costs (% sales)	0.9	11.5	5.1
Overstock costs (% sales)	15.6	4.7	11.6
Total overstock and omit costs (% sales)	16.5	16.2	16.7

While the numbers in this example are specific to a particular line of dresses at one retailer, the problems highlighted in this example are symptomatic to this industry. Markdowns as a percentage of sales since World War II have increased from 3 to 16% for many segments of the apparel industry as described in Pashigan (1988). A Survey conducted by a major retailer which was reported in the New York Times on June 2, 1994 concluded that 50% of customers did not purchase products when they visited a store and of these 40% stated that they did so because they did not find what they specifically wanted. Similar over and under supply costs have been observed for products in other industries like automobiles (White, 1992) and personal computers (Hooper and Yamada, 1992; Stewart, 1992).

To understand why these costs are prevalent, one must look for the root causes. Overstocking and stockout costs arise due to the inability to predict and rapidly react to customer preferences. Changes in the retail environment like consolidation and increasing product variety have resulted in large chain stores offering large varieties of merchandise, making it difficult to understand preferences. Global sourcing with its associated longer lead times has constrained the ability to rapidly react to changing customer preferences. Scaling back on these changes may not be viable as the benefits both in the cost and revenue sides far outweigh the increased costs due to these supply demand mismatches. Retailers realize that reducing these costs even by a small percentage could significantly increase profits. However, achieving this reduction in costs requires an intellectual transition from right brained or artistic thinking to left brained or analytical thinking. Left brained thinking has to be supported by a solid analytical framework designed to improve the accuracy of the merchandise planning process.

In this chapter, we develop optimization models for two key aspects of the merchandise planning process. In the first aspect we consider the process commonly used by retailers to gauge the salability of new products known as merchandise testing. In the second aspect we consider replenishment based on actual sales, a strategy used by retailers to minimize the inventory risk associated with an assortment. For each of these aspects, we precisely define the problem and its associated issues. We then develop analytical models to improve the accuracy of the associated decision process. To evaluate our techniques we compare their performance to existing practices using actual sales data in real retail environments. Finally we develop general insights to improve the efficiency of these aspects and discuss possible extensions of our methods.

The remainder of this chapter is organized as follows. In Sect. 2 we discuss merchandise testing. In Sect. 3, we analyze replenishment planning. In the concluding chapter, we suggest provide some important conclusions and suggest areas for possible future research for each of these aspects.

2. Merchandise Testing[24]

2.1 Introduction

Retailers segment merchandise into basic and fashion products. Basic products have relatively stable demand and a long life cycle, which makes it fairly easy to forecast demand and manage inventory for a particular product using standard methods that rely on a sales history of the product. Forecasting and inventory management are much harder for fashion products. Their demand is highly unpredictable and they have a short life cycle, typically just a few months. They are often bought just once, at a time prior to the start of the actual sales season and the decision of how much to buy is not based on actual sales of the product, but merely on the subjective judgement of merchandisers and buyers on how well it will sell. We have found that these subjective forecasts typically have an average error of 50%

---

[24] This section has been adapted from Fisher and Rajaram (2000).

or more. As a result, retailers frequently buy too little of some fashion products, resulting in lost sales and profit margin, and too much of other products, resulting in excess supply that must be marked down in price at the end of the season, frequently to the point where the product is sold at a loss.

To reduce these costly forecast errors, many retailers conduct experiments called tests in which products are offered for sale under carefully controlled conditions in a small number of stores. Tests are used to measure consumer reaction to a variety of variables including price, floor placement, marketing message or some aspect of styling such as color, fabric or silhouette. We focus here on a particular type of test called a depth test used to predict the season sales of a particular product. In a depth test, a supply of the product sufficient to avoid stock-outs is placed in a small sample of stores for a 2- to 3-week period just prior to the start of the regular sales season. Sales in the test stores during this period are used to predict season sales for the chain and this forecast is used as a basis for initial or replenishment orders. Although we focus specifically on depth testing in this section, our method for choosing test stores could also be useful in other types of testing. For simplicity, we'll refer to this type of test in this section as a merchandise test, or simply a test.

A retailer faces several issues in designing an effective merchandise test, including how many and which specific stores in which to conduct the test and how to create a forecast for the entire chain based on test store sales. The decision of how many test stores to use must tradeoff the increased accuracy that comes from using more test stores against the cost of running the test, which is greater if more stores are used. The cost of running a test results from administrative costs, the need to provide extra inventory to avoid stock-outs during the test, possibly the cost of air-freighting merchandise to the test stores, and an opportunity cost on the store space used for the test resulting from the fact that test merchandise by its nature usually sells less well on average than regular merchandise. The high cost of testing generally leads retailers to use a small number of test stores (e.g., 5–25). Choosing a small sample of test stores from the hundreds of stores that comprise a large chain is challenging due to the typically large variation in store characteristics such as location, climate, size and demographics of the surrounding customer base.

Despite the practical relevance and complexity of this problem, we found nothing in the academic or managerial literature that describes how to design an effective merchandise test. There is extensive academic literature on test marketing (e.g., see Urban and Hauser (1980)) that would appear to be relevant, but turns out not to be directly applicable since they involve longer duration and observation of trial and assume repeat purchases.

A number of articles that review current retail practice (Doyle and Gidengil (1977), Fox (1995), Hollander (1986), Pollack (1994) and Wilson et al. (1995)) emphasize the importance of merchandise testing and highlight a need for more effective procedures, but do not themselves describe how to conduct an effective merchandise test. Doyle and Gidengil (1977) review merchandise testing as part of the broader topic of retail experimentation and conclude that these methods, despite enormous potential, have thus far "made little contribution" to retailing because of problems both practical and theoretical.

We have been able to gather systematic information on testing practice as part of a broader multi-year project involving 32 leading retailers of fashion type products, including apparel, computers, consumer electronics, entertainment software, books, music, toys, watches and jewelry (See Fisher et al. (2000) for more details). Of the 27 retailers who answered the questions on testing, 25 indicated that they conducted merchandise tests of some type, which supports that testing is widely used by retailers. Retailers were also asked to rank the effectiveness of their testing program on a 10 point scale, defining a "10" as the ability to predict sales from a test with an error of about 10%. The median answer was 6, suggesting that there is considerable room for improvement in testing accuracy in practice. In follow-up interviews, it appeared that the retailers who tested best had done an excellent job on many of the practical issues of testing, but had not used any statistical methods to determine the most representative stores to use for a test or to interpret the results of a test. While retailers had diverse theories on how to test, to our knowledge none of them have subjected alternative approaches to a rigorous comparison to determine what actually works best.

This section presents a methodology for resolving two key decisions in designing a merchandise test: which stores in which to conduct the test and how to create a season forecast for the chain based on test results. To choose test stores, we cluster the stores of the chain into groups that are similar based on historical sales of products similar to those to be tested. We apply regression to this sales history to estimate a linear model to predict season sales for the chain from test store sales. We also consider clustering based on various store descriptor variables. Using data from three retailers – a specialty apparel retailer and two shoe retailers, Nine West and Meldisco, we compare the clustering methods to the existing process used by the apparel retailer and to two standard statistical approaches based on forecast accuracy and the cost of a supply plan guided the forecast. We find that sales based clustering significantly outperform the other methods on both of these metrics.

Next, we present the basic ideas behind our methodology. Section 2.3 presents the optimization models used to form clusters and select a test store within each cluster and predict total season sales across all the stores of the chain from test store sales. Section 2.4 reports an application of these ideas to a retailer specializing in women's fashion apparel, with over 1,000 stores nationwide. Compared to the current testing process in place at the retailer, our method would reduce the cost of stockouts and of merchandise left over at the end of the season by enough to increase profit by 100%.

2.2 Methodology

We describe a methodology for resolving the key decisions in designing a merchandise test: how many and which specific stores in which to conduct the test and how to create a season forecast for the entire chain based on test store sales. We assume the retailer has (1) identified a set of products within a classification that they would like to test, (2) specified the time interval within which the products will be sold (the sales season), and (3) determined a test period during which the products will be offered for sale in selected stores to test their sales potential.

As mentioned in the introduction, retailers typically use a test period of 2–3 weeks' just prior to the start of the regular season. The duration of the test presents the same tradeoff as how many stores to test in: a longer test is more accurate, but more expensive. We have found 2–3 weeks to be typical practice, which seems to make sense. The test needs to be at least 1 week to control for inter-week temporal effects. If the test occurs just prior to the start of the regular season, then the ending point of the test is constrained by the need to position supply based on the test by the start of the regular season. Hence, a longer test must start earlier and the earlier the test begins, the greater the difference in conditions between the test period and the period when products will be sold.

We define the primary sales season as the period within the sales season during which the selling price of the product exceeds acquisition cost plus variable selling expenses, and the level of inventory at each store is sufficient to prevent supply shortages. We assume that the same price is charged at all stores during this period, although this common price may vary over time. It is important to restrict the sales data to the primary sales season; otherwise, sales below cost and supply shortages, both outcomes of bad planning, could distort the distribution of sales at individual stores. Note that this also insures that sample sales are not influenced by substitution due to a stock-out of the product the customer was seeking. However, sample sales can be influenced by the dependency of demand with complementary or competitive products. If a store contains more complementary (competitive) products during the test than will be there during the regular selling season, then this will bias test sales upward (downward). This is an inevitable source of noise in a merchandise test, but it can be minimized by making the set of products offered during the test period as similar as possible to what will be offered during the regular season.

Since the purpose of merchandise testing is to create a forecast used to determine purchase quantities, it is important to understand the costs that result from forecast errors. If S_p is the actual demand for a product p during the primary season, U_p the per unit cost of buying less than demanded and O_p the per unit cost of buying more than demanded, then the cost associated with forecast $\hat{S}_p$ is $U_p \max(S_p - \hat{S}_p, 0) + O_p \max(\hat{S}_p - S_p, 0)$. The understock cost U_p is often taken to be the profit contribution margin (price minus variable cost) that is lost if there is insufficient supply to meet demand. Excess supply is usually marked down in price at the end of the season and sold at a loss, so O_p is set to variable cost minus the marked down price.

Before describing our computational procedure, we believe that it will be helpful to outline some features of retailing considered in designing our method. We can think of a product as being defined by a set of values for various attributes. Consumers differ in their preference for attribute values, and hence the same product will have different appeal to different customers. In a retail chain with a large number of stores, consumer preferences will differ from store to store. While it is hard to directly measure the attribute preferences of customers that shop a given store, given that attribute preference influences purchase decisions, the actual sales of a store can be thought of as a summary of the attribute preferences of customers at that store. In particular, if the percentage mix of products bought by customers at two different stores is similar, then we might infer that the customers of the two stores have similar preferences. This is the whole basis for micro-merchandising, a practice followed by a significant number of retailers in which a unique assortment of merchandise is offered in each store (or a group of similar stores) tuned to maximize the appeal to customers of that store (see Patterson (1995); DiRomualdo (1998) for examples).

Our approach to testing is designed to recognize these features by using past sales of similar products to identify a set of test stores that collectively span the diverse segments of a large chain. In Sect. 2.3, we first describe a procedure for choosing test stores assuming that the number of test stores k has been specified. Then, we show how to set k to minimize the combined cost of testing and of forecast errors resulting from the test.

2.3   Model Description

We assume that the retailer has assembled a sales history of comparable products that were offered in a prior season or seasons. Appropriate comparable products are usually last year's products within the same classification. Let n denote the number of stores in the chain, m the number of previous products for which we have a sales history, S_{ip} the observed sales during the primary sales season at store i for product $p, S_p = \sum_{i=1}^{n} S_{ip}$, and $\overline{S}_{ip}$ the sales of product p in store i during a period comparable in timing and duration to a period during which a test would be conducted.

To choose k test stores, we partition the n stores of the chain into k clusters. The stores within each cluster are chosen to minimize a measure of dissimilarity based on the percent of total sales represented by sales of each of the prior products in each store. Two stores that sold exactly the same percentage of each of the prior products would be in the same cluster, and all of the stores within a cluster would sell approximately the same percentage of each of the prior products. We then choose a single test store within each cluster that best represents the cluster in the sense that using test sales at this store to forecast sales of other stores in the cluster minimizes the cost of forecast errors.

We first describe the optimization model used to form clusters and select a test store within each cluster. This model is a specialized integer program known as the k-median problem, which we solve with the highly efficient algorithm given in Cornuejols et al. (1977).

Variables

$$y_j = \begin{cases} 1, & \text{if store j is chosen as a test store} \\ 0, & \text{otherwise} \end{cases}$$

$$x_{ij} = \begin{cases} 1, & \text{if store i is assigned to a cluster represented by test store j} \\ 0, & \text{otherwise} \end{cases}$$

Parameters

$I = (1, \ldots n) = $ store index set

$P = (1, \ldots m) = $ prior product index set

$$w_i = \sum_{p \in P} S_{ip}$$

$$\beta_{ip} = \frac{S_{ip}}{\sum_{p \in P} S_{ip}} , \ i \in I, \ p \in P$$

$$d_{ij} = \sum_{p \in P} (U_p \max(\beta_{ip} - \beta_{jp}, 0) + O_p \max(\beta_{jp} - \beta_{ip}, 0))$$

The Test Store Selection Problem (TSSP) is represented by the following integer program.

$$\text{Min} \sum_{i \in I} \sum_{j \in I} w_i d_{ij} x_{ij} \tag{1}$$

Subject to:

$$\sum_{i \in I} x_{ij} = 1, \ j \in I \tag{2}$$

$$\sum_{j \in I} y_j = k \tag{3}$$

$$0 \le x_{ij} \le y_j \le 1, \quad i, j \in I \tag{4}$$

$$x_{ij} \text{ and } y_j \text{ integral}, \ i, j \in I \tag{5}$$

Equation (2) enforces the condition that each store is assigned to a test store, (3) that we have exactly k test stores, and (4) that stores are only assigned to chosen test stores. Objective function (1) is structured to select k test stores that minimize the total overstock and understock costs if total test store sales are extrapolated to develop forecasts for stores assigned to it. To illustrate this point, let store i be represented by test store j (i.e. $x_{ij} = 1$). Then it would not be unreasonable to forecast total sales for store i as $\frac{w_i}{w_j} S_j$ and sales for individual products in store i as $\frac{w_i}{w_j} S_{jp}$. The total cost associated with this forecast at this store is

$$\sum_{p=1}^{m} U_p \max(\frac{w_i}{w_j} S_{jp} - S_{ip}, 0) + O_p \max(S_{ip} - \frac{w_i}{w_j} S_{jp}, 0)$$

$$= \sum_{p=1}^{m} w_i (U_p \max(\frac{S_{jp}}{w_j} - \frac{S_{ip}}{w_i}, 0) + O_p \max(\frac{S_{ip}}{w_i} - \frac{S_{jp}}{w_j}, 0))$$

$$= w_i d_{ij}$$

Hence, the k-median problem can be interpreted as forming clusters and choosing a test store in each cluster that minimizes the cost of forecast errors if total season sales at the test store were the predictor variable for its cluster. This also minimizes the expected cost of forecast errors based on test period sales provided the percentage of season sales occurring during the test period is approximately the same for all stores within a cluster. In this regard, it is worth noting that stores can differ not only in their sales mix, but in the timing of their sales. If the timing of sales differs, then two stores might have an identical sales mix for the season, but a different mix during the test period. In this instance, one of the stores might not be a good predictor for the other. As a practical matter, we have found that the most common cause of timing differences is climate. Southern stores sell spring/summer merchandise earlier than northern stores, and fall/winter merchandise later. We found in our testing that these climate differences also cause a difference in sales mix. Hence, while a timing difference that occurs separately from a mix difference is a potential problem, it did not occur in our test data. Were this to be an issue, we would recommend redefining d_{ij} to be based on the difference in the mix of season sales for store i and test period sales for store j or to use a combined optimization approach defined in the Appendix.

We also tested a version of our method in which the d_{ij} were set based on a weighted combination of several store descriptors rather than on differences in stores' sales mix. The store descriptors were store latitude and longitude (the distance between two stores on this measure was simply the Euclidean distance between the stores), average temperature during the sales season, total store sales, ethnicity (percentage of white in the postal code where the store is located – not politically correct, but believed by many retailers to be indicative of sales patterns), neighborhood type (either urban, suburban or rural – we assigned a distance of 0 if the neighborhoods were the same and 1 if they were different) and store type (either mall, strip mall or college campus – we assigned a distance of 0 if the store types were the same and 1 if different). The distance between two stores used in clustering was a weighted combination of the absolute difference of these six measures, with non-negative weights summing to one chosen to equalize the average influence of each of the six measures. We also tested a combined approach in which the distance between two stores was based on a combination of the six store descriptors and the difference in stores sales mix, with weights chosen to equalize the impact of these seven factors.

While we suggested above that w_i/w_j was a reasonable weight to use in predicting sales of a product at store i from it's sales at store j, to insure that we have the best possible weights and to adjust for the fact that test sales are for a shorter time interval than the full season, we use the following linear program, which we call the Test Sales Extrapolation Problem (TSEP), to determine the optimal set of weights ($\alpha_1 \cdot \alpha_k$), which scales test sales to estimate product sales for the entire season across the chain for a given set of k test stores.

$$Z_1(k) = \text{Min} \sum_{p \in P} U_p \theta_p + O_p \gamma_p$$

$$\theta_p \geq \hat{S}_p - \overline{S}_p \quad \forall p \in P$$

$$\gamma_p \geq \overline{S}_p - \hat{S}_p \quad \forall p \in P$$

$$\hat{S}_p = \sum_{i=1}^{k} \alpha_i \overline{S}_{ip} \quad \forall p \in P$$

$$\theta_p, \gamma_p \geq 0 \quad \forall p \in P$$

In the discussion so far, we have assumed that the number of test stores k is given. To find the best number of test stores, we would choose k to minimize $C(k) = Z_1(k) + C_T k$, where C_T is the fixed cost for testing at each store. The fixed cost C_T results from the administrative cost of running the test at a store and the fact that test merchandise by its nature usually sells less well on average than regular merchandise, and hence there is an opportunity loss on the shelf space dedicated to the test. Our procedure is fast enough that we can solve the optimization on k by enumeration of all values k = 1, 2, ... noting that the largest number of test stores we need consider is bounded by $C(k)/C_T$ for any value of k for which we have evaluated C(k), since using a number of stores larger than this would incur fixed costs greater than the total cost of a known solution.

It is possible to formulate a single optimization model to choose the identity of test stores and the weights in the linear forecast formula so as to minimize the total cost of over-stock, under-stock and testing. We tested this unified approach on the data described in Sect. 2.4 and found that it performed slightly worse than the approach described above. Further details of the unified approach are provided in the Appendix, including the formulation and our computational experience with it.

In the final analysis, the real test of our method is its ability to improve the accuracy of the merchandise testing process in an actual retail environment, a question we consider next.

2.4   Application at a Women's Specialty Apparel Retailer

We have tested these ideas with a large specialty women's apparel retailer with net sales in 1993 exceeding $1 billion and with more than 1,000 stores dispersed throughout the U.S. In addition to varying by region, stores vary by size, format (mall, strip mall, etc.), urban vs. suburban and the ethnicity of their target customer base. This retailer relied on testing as their primary tool for forecasting the demand for new products, but they also believed that the accuracy of their current methodology could be significantly improved, which is what led to our involvement.

In developing merchandise plans, the retailer divides the year into two seasons: fall/winter and spring/summer. The first week of the fall/winter season is the first full week of September, while the first week of the spring/summer season is the first full week in April. In their current methodology, they select 25 test stores whose total dollar sales are close to average store sales for the chain. Product tests are conducted at these stores over a 3-week period. To develop a season forecast and supply plan for the entire chain, total sales during the test period are divided by two factors estimated from past sales history. The first factor equals the proportion of season sales that are historically observed during these 3 weeks, while the second factor equals the proportion of total sales that are observed at these 25 stores.

To evaluate our approach and compare it to this approach, we considered the women's knit tops division, which historically represents one of the single largest portions of investment and the highest level of sales uncertainty. The data available on which to evaluate our approach consisted of 1993 sales by store, by week and by size of the 250 style/colors that comprise the products of this division. To estimate the timing and duration of the primary sales season in this division, we analyzed 1993 sales data at the store/size level for these 250 style/colors. This analysis revealed that merchandise was in place at all stores by the start of week 3, that there was sufficient inventory at all the stores to meet potential demand through week 12, and that products were sold below cost typically after week 18. Consequently, we used a 10-week period from the beginning of week 3 through the end of week 12 as the primary sales season.

We used these data to simulate the way our method would have performed if it had been used to design a test of this merchandise to be conducted during weeks 3 through 5. We used half of the 250 style/colors to fit the parameters of our model and the other half to test the accuracy of the model predictions. The planners at the knit top division classified the 250 style/colors into 16 product groups based on similarity in style and fabric texture. We selected 125 style/colors by choosing half the products in each of these groups. Let P denote this set of selected styles and $\overline{P}$ the remainder. For each $p \in P$, we used the sales data by store and product over the 10-week primary season to calculate w_i and d_{ij} and solved (TSSP) to optimality by the technique described in Cornuejols et al. (1977). For a given k, assume without loss of generality that the chosen test stores are indexed 1 through k. We then solved (TSEP) using the OSL solver in GAMS (see Brooke et al. (1992)) to develop the linear function $\hat{S}_p = \sum_{i=1}^{k} \alpha_i \overline{S}_{ip}$, where $\overline{S}_{ip}$ is the actual sales of product p at store i during weeks 3 through 5, the period when a test would be conducted. For each $p \in \overline{P}$, we computed $\hat{S}_p = \sum_{i=1}^{k} \alpha_i \overline{S}_{ip}$, using $\overline{S}_{ip}$, the actual sales of product p at store i during week 3 through 5. We estimated the forecast error $\sum_{p \in \overline{P}} \left| \hat{S}_p - S_p \right|$, where S_p is the actual sales of product p across the chain during the primary season. Based on average selling price and costs during this period, we calculated the cost of these errors as the loss due to selling below cost and lost margin due to supply shortage, namely $\sum_{p \in \overline{P}} U_p \max(S_p - \hat{S}_p, 0) + O_p \max(\hat{S}_p - S_p, 0)$. We also performed these computations for the version of our method in which d_{ij} is determined from a composite of store descriptor variables as described in the preceding section.

To provide a comparison, we also evaluated forecast errors and cost for the rules used by the planners at this retailer, the k median method based on store descriptors, alone and combined with sales mix differences, and two standard approaches to variable selection in linear regression, since the problem of choosing k test stores and a linear prediction function based on test sales at these stores can be viewed as choosing the best k out of n possible variables in a linear regression. Given actual sales S_p and test sales $\overline{S}_{ip}$ for $i = 1, \ldots n$ and $p = 1, \ldots m$, we used the forward selection and backward elimination methods (Myers (1990)) to choose k out of the n test store sales variables that best predict actual sales in the sense of minimizing the coefficient of determination R^2.

Notice that our method provides not only a sales forecast for the chain, but for each cluster. The cluster forecasts were used to guide allocation of product to stores for the k-median clustering based on sales method. In our prediction formula, $\alpha_p S_{pj}$ represents the forecast of product j in the cluster of stores corresponding to test store p, and hence this quantity is the ideal amount to send to this cluster. Total sales volumes at individual stores were used as a basis to determine allocations to stores within a cluster through the formula $\alpha_p \overline{S}_{pj} w_i / \sum_{i \in I_p} w_i$. For other methods, we used the existing approach of the retailer of allocating total supply in proportion to historical sales volume.

Forecast errors and costs for all six approaches for different values of k are shown in Table 2. Forecast errors and costs are an aggregate of these values determined at the store/style/color level. In aggregating forecast errors, we summed over all stores, styles and colors the absolute difference between forecasted and actual sales and expressed it as a percentage of total actual sales. These results show that the k-median approach with clustering based on sales is significantly more accurate than the other methods. For example, for k = 10, it reduces costs due to forecast errors relative to the regression methods, the existing method in use at the retailer or clustering based on store descriptors by at least 8% of revenue. This improvement drops straight to the bottom line and would more than double profit when one considers that retailers typically earn profit before income tax of about 3–5% of sales. The combined approach of clustering on store descriptors and sales mix difference comes close to, but is still dominated by clustering on sales alone.

Table 2. Forecast error (F.E) as a % of unit sales and costs as % of revenue for different methods

K	k-Median clustering based on sales		Regression with forward selection		Regression with backward estimation		Existing method used by retailer		k-Median clustering based on store descriptions		k-Median clustering based on store descriptions and sales	
	F.E.	Costs	F.E.	Costs	F.E.	Costs	F.E.	Costs	F.E.	Costs	F.E.	Costs
1	36	40	37.9	42.2	38.6	43	71	74	39.8	44.3	38.2	42.4
5	17	25.9	27	35.9	28	37	52.1	58	23.3	32.8	20.3	30.6
10	12.9	20.9	19.1	29.2	20	30.1	41.9	46	19.0	28.9	16.7	23.0
15	12	19.5	17.1	26	16.8	23.7	38.1	42.3	16.8	23.2	15.9	22.4
16	11.7	19.3	16.8	23.6	16.6	23	37.7	41.9	16.5	22.9	15.4	22.2

We believe one reason our method outperforms the forward selection and backward elimination methods is that these methods seek to minimize squared errors while our method optimizes the true cost of forecast errors. In addition, our approach, which is based purely on sales, outperforms descriptor variables as it is not always clear which are the best store descriptors and how best to combine them. However, the sales based process is completely objective and directly corresponds to the retailer's objective of minimizing the understock and overstock costs of forecast error.

To better understand the underlying causes of sales differences, it may be desirable to relate sales differences as much as possible back to store descriptors. To do this, we correlated store distance measures based on sales differences with the store descriptor variables temperature, ethnicity, location, store type and size. We find these descriptors as a group explain 85–89% of the variation in sales differences, with temperature being by far the most significant variable. We also found that, contrary to popular belief, store size and location had little impact on sales patterns. Details of this analysis can be found in Fisher and Rajaram (2000).

We also conducted extensive analyses to address the following issues (details of can also be found in Fisher and Rajaram (2000): (1) Assessing how the degree of collinearity between sales at the test stores affects the performance of these methods; (2) Justifying our choice of ten test stores to perform this analysis and evaluating the sensitivity of our method to the choice of the specific test store within a cluster; (3) The relationship between store size and test stores; (4) The effect of temperature as an explanatory variable in the formation of the clusters.

Having evaluated our method on 1993 sales data, we also wanted to determine how well it would perform across multiple years, since in actual use we would be fitting the model on sales that had occurred 1-year prior to actual introduction of the styles being tested. We obtained sales data for 30 style/colors from the 1994 fall season in the Knit Tops division that had been tested in 25 stores chosen by the planners. These were all new products that had not been on sale during 1993. We fit our clustering model as fit on the 1993 data to these 30, 1994 products. This exactly replicates how the model would be used in practice and hence is an accurate measure of its effectiveness.

Using the actual primary season sales (S_q) for these products in 1994, we computed total forecast error $\sum_{q \in Q} \left| \hat{S}_q - S_q \right|$ for all forecasts. Based on cost and selling price for these products during each week of 1994, we computed for all supply plans, the loss due to selling below cost and lost margin due to supply shortages. Forecast errors and costs for all plans are shown in Table 3. The reduction in cost achieved by our plan relative to the other methods is

6.5–18%. This is better appreciated if we consider the fact that the recorded net before tax profit in 1994 for these styles was 9% of revenue. Hence, our method could potentially improve profits by from 6.5/9=72% to 18/9 = 200%.

The way we analyzed costs for these 30 style/colors treats demand as sales. However, because of stockouts, true demand usually exceeds sales. This generally biased comparisons against our method relative to the existing method since the inventory levels that conditioned sales had been determined by the existing method.

Table 3. Forecast error and cost for models fit on 1993 data and applied to 30, 1994 products

	Existing method with 25 test stores	k–Median clustering based on sales with 10 test stores	Regression with forward selection with 10 test stores	Regression with backward elimination with 10 test stores	k–Median clustering based on store descriptors with 10 test stores	k–Median clustering based on store descriptors and sales with 10 test stores
Forecast error as % of sales	38	19.5	30	31	29	25.5
Markdown cost as % of revenue	30.6	19	28	29	25	24
Lost margin as % of revenue	14.4	8	11	12	11	9.5
Total of markdown cost and lost margin	45	27	39	41	36	33.5

Assessing this effect is difficult as information on lost demand is not recorded. To estimate lost demand and margins, we first identified when every product/size combination in this group sold out in each store. We then defined the profitable season as the period during which products sold above cost and tabulated the percentage of profitable season sales that occurred each week for this group in total. This data was used to estimate the sales that would have occurred after a stockout in a store/product/size combination before the end of the profitable season.

We applied our method for estimating lost demand to 104 style/color/size combinations in the knit seris group and estimated that in total, lost demand due to stockouts equaled 134% of sales actually realized. The lost margins due to this potential 134% increase in sales were around 57% of sales actually realized. Even if we allow for substitution between products to lower these estimates, the lost margins are clearly enormous.

Much of this lost demand was due to inaccurate distribution of inventory across sizes. Often, we observed that a particular size in a given style/color accounted for a large proportion of stockouts, while other sizes of this style/color had to be eventually sold at markdowns below cost. To reduce such misallocation within the size distribution of a product, our clustering method could be applied to historical data on sales by size for a product to form clusters of products that had similar size selling patterns. We would then examine the nature of the products in each cluster as a way to understand how size distribution differs by product. For example, this might result in ten distinct size distributions and a definition of the types of products that had a particular distribution. A new product could be assigned the size distribution of the product type that it best fits.

Recall that a test period of 3 weeks was used in developing the data reported in Table 4. To determine the impact of the length of the test period on cost and forecast error we applied k-median clustering based on sales with test periods of varying lengths. Results are reported in Table 4 and suggest the industry practice of a 3-week test period is not unreasonable.

Table 4. Impact of test period length on forecast error and cost for k-median clustering based on sales

Length of test period	Forecast error as % of sales	Total markdown cost and lost margin as % revenue
1	30	42
2	24	33
3	19.5	27
5	17	24
7	16	23.2
10	15.5	22.7

3.   Optimizing Inventory Replenishment[25]

3.1   Introduction

Retail inventory management is concerned with determining the amount and timing of receipts to inventory of a particular product at a retail location. Retail inventory management problems can be usefully segmented based on the ratio of the product's life cycle T to the replenishment lead-time L. If T/L < 1, then only a single receipt to inventory is possible at the start of the sales season. This is the case considered in the well-known newsvendor problem. At the other extreme, if T/L >> 1, then it's possible to assemble sufficient demand history to estimate the probability density function of demand and to apply one of several well-known approaches such as the Q,R model.

The middle case, where T/L > 1 but is sufficiently small to allow only a single replenishment or a small number of replenishments, has received much less attention both in the research literature and in retail practice. As we will describe in Sect. 2, there is a small, but growing literature on limited replenishment inventory problems. Perhaps because of the lack of published analysis tools, we have found that retailers often ignore the opportunity to replenish when T/L is close to 1 and treat this case as though it were a newsvendor problem. This is unfortunate, since, as we'll show in the numeric computations in this section, planning for even a single replenishment in this case can dramatically increase profitability.

In this section, we consider limited life-cycle retail products in which only a single replenishment is possible. We model as a two-stage stochastic dynamic program the problem of determining the initial and replenishment order quantities so as to minimize the cost of lost sales, backorders and obsolete inventory at the end of the products life. We show that the second stage cost function of this program may not be convex or concave in the inventory position after the reorder is placed, which means that simulation based optimization techniques (Ermoliv and Wetts, 1998) typically used to solve problems of this type are not guaranteed to find an optimal solution. For this reason, and also for computational efficiency, we formulate a heuristic for this problem. We show that this heuristic finds an optimal solution if demand subsequent to the time at which a reorder is placed is perfectly correlated with demand prior to this time. While perfect correlation between early and late demand is unlikely, we believe this result indicates that our heuristic will work well if this correlation is high. Thus, in practice, it seems reasonable to expect good performance from this heuristic since the

---

[25] This section has been adapted from Fisher, Rajaram and Raman (2001).

logical basis of implementing replenishment based on early sales is that demand during the later season is highly correlated with early demand. In our application, the correlation between early and late demand was 0.95. We also apply a simulation based optimization techniques (Ermoliv and Wetts, 1998) and find that our the heuristic is much faster and finds solutions within 1% of the optimization procedure if the correlation between early and late demand is at least 60%. For lower correlations, the solutions are within 1–5%.

We have applied this process at a catalog retailer and find that it improved over their current process for determining initial and replenishment quantities by enough to essentially double profits. Remarkably, compared to no replenishment, a single optimized replenishment improves profit by a factor of five. A key challenge in implementing short life-cycle replenishment is estimating a probability density function for demand with no demand history. To circumvent this problem in our application, we applied the committee-forecast process in Fisher and Raman (1996) and found that it worked well.

The most important difference between catalog and traditional retail management is that a catalog customer will generally accept a backorder if an item is stocked out. Because our application was at a catalog retailer, our model and heuristic are given for this version of the problem, but it is straight forward to modify the model, heuristic and proof of optimality for the case where backorders are not allowed.

In Sect. 3.2, we review the literature on short life-cycle inventory replenishment. In Sect. 3.3, we formulate the problem; in Sect. 3.4, we state our heuristic and establish optimality conditions; in Sect. 3.5, we show how to modify the process when customers may return merchandise and in Sect. 3.6, report results on our application.

3.2 Literature Review

Analytical models for managing inventory for short life-cycle products share many common features. First, all are stochastic models, as they consider demand uncertainty explicitly. Second, they consider a finite selling period at the end of which unsold inventory is marked-down in price and sold at a loss. In this sense, these models are similar to the classic news-vendor model. Third, they model multiple production commitments such that sales information is obtained and used to update demand forecasts between planning periods. The last two characteristics: finite selling periods and multiple production commitments differentiate style goods inventory models from other stochastic inventory models. Examples of papers that consider style good inventory problems include Bitran et al. (1986), Fisher and Raman (1996), Hausman and Peterson (1972), Matsuo (1990) and

Murray and Silver (1966). A detailed review of these papers can be found in Raman (1999).

Recent work that deals specifically with the retailer's inventory management problem for short life-cycle products are Bradford and Sugrue (1990) and Eppen and Iyer (1997a, 1997b). Bradford and Sugrue model a decision that is similar to the one we study, but do not consider the impact of replenishment lead-times. In addition, their solution procedure consists of complete enumeration, which works efficiently for smaller problems but could be difficult to implement in larger, practical sized-problems. Eppen and Iyer (1997a) consider a problem that is substantially different from ours. Even though their model allows the retailer to "buy" and "dump" at the beginning of each period, the solution method they propose applies only when no "buy" decisions are permitted after the first period.

Eppen and Iyer (1997b) model a backup agreement in place at a catalog retailer. A backup agreement is one of the mechanisms by which a retailer achieves replenishment of branded merchandise supplied by a manufacturer to several retailers. In a backup agreement, a retailer places an initial order before the start of the sales season and commits to reorder a certain quantity during the season. After assessing sales during the early part of the season, if the retailer chooses to reorder less than this commitment, there is a penalty cost assessed for each unit not ordered. In this model, replenishment lead-times are assumed to be zero, which is reasonable since the manufacturer would typically have produced the product and be holding it in inventory for this and other retailers. Because the replenishment lead-time is zero, it is not necessary for the retailer to accept backorders from consumers.

In this section, we consider the case where replenishment is achieved without backup agreements. After receiving an updated order, the manufacturer produces and delivers products to the retailer after a significant lead-time. The retailer is not required to commit to any of the reorders. To compensate for the long lead-time, consumer backorders are accepted by the retailer. This case occurs when the manufacturers are either captive suppliers or wholly-owned by the retailer and the retailer sources from several such manufacturers. Thus in this case, it is crucial to model the impact of lead-times and backorders, although this significantly complicates the analysis leading to a non-convex optimization model. In addition to incorporating replenishment lead-times and backorders, our work differs from all these papers in the process that we use to estimate demand densities and to compare our method to actual practice.

3.3 Model

We model the supply decisions faced by a catalog retailer for a product with random demand over a sales season of fixed length. The retailer must determine an initial order Q_1 available at the start of the sales season. At a fixed time t during the season, the retailer updates the demand forecast based on observed sales and places a reorder quantity Q_2 that arrives after a fixed lead-time L at time $t + L$.

Price is fixed throughout the season. Inventory left over at the end of the season is sold at a salvage price below cost. Customers who encounter a stockout will backorder if there will be sufficient supply at some point in the future to satisfy the backorder. Specifically, the opportunity to backorder is not offered to a customer once the total supply quantity $(Q_1 + Q_2)$ has been committed through sales or prior backorders. A lost sale is incurred when an item requested by a customer is not in stock or not backordered.

We first model this problem and formulate a solution heuristic assuming that the reorder time t is fixed. Then, we determine an optimal reorder time t empirically for a given data set by parametrically solving this problem with varying t. We are given:

C_u: Cost per unit of lost sale. This is set to the difference between the per unit sales price and cost of the product.

C_o: Cost per unit of left over inventory. This is set to the difference between the per unit cost and the salvage price of the product.

C_b: Cost per unit of backorders. This is set to the additional costs incurred in procurement and distribution when an order is backlogged plus an estimate of the cost of customer ill will.

L: Length of replenishment lead-time.

Define the following variables:

X: Random variable representing total demand until the reorder is placed.

Y: Random variable representing total demand during the replenishment lead-time.

W: Random variable representing total demand *after* the reorder arrives until the end of the season.

R: Random variable representing total demand after the reorder *is placed* until the end of the season, where $R = Y + W$.

Q_1: Initial order quantity

Q_2: Reorder quantity

I: Inventory position after reorder is placed. $I = Q_1 + Q_2 - x$.

For the given reorder time t, the decision process involves choosing Q_1, observing x and then determining the inventory position I for the remainder of the season to minimize total backorder, understock and overstock costs. This sequence of decisions is shown in Fig. 1.

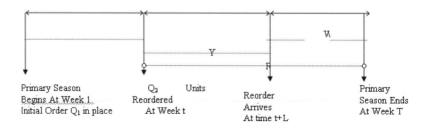

Figure 1. The replenishment planning process

We consider random variables Θ and Ψ with joint density function $f(\theta,\psi)$. Let $g(\theta)$ be the marginal density on Θ defined by $f(\theta,\psi)$ and $h(\psi|\theta)$ be the conditional density on ψ given θ defined by $f(\theta, \psi)$. We define $E_{\Psi/\Theta}(\Omega(\psi)) = \int_0^\infty \Omega(\psi)h(\psi\,|\,\theta)\partial\psi$ and $E_{\Theta}(\Omega(\theta)) = \int_0^\infty \Omega(\theta)g(\theta)\partial\theta$, where $\Omega(\ )$ is any real valued scalar function. Let $(a)^+ = \max(a,0)$. We model this problem by the two stage stochastic dynamic program (P1).

$$Z(t) = \min_{Q_1 \geq 0} C(Q_1) = E_X\,[C_1(Q_1,x) + C_2(Q_1,x)] \tag{P1}$$

Where,
$$C_1(Q_1,x) = C_b\,(x-Q_1)^+$$

$$C_2(Q_1,x) = \min_I C_2(I,x)$$
$$= E_{Y/X}\{C_b \min((y-(Q_1-x)^+)^+, I-(Q_1-x)^+)\} \tag{P2}$$
$$+ E_{Y/X}E_{W/X}\{C_u(y+w-I)^+ + C_o(I-y-w)^+\}$$

$$I \geq Q_1 - x$$
$$I \geq 0$$

$C_1(Q_1,x)$ represents the backorder costs $C_b(x-Q_1)^+$ during the period before the reorder is placed. $C_2(I,x)$ represents expected cost as a function of the inventory position I after the reorder is placed and consists of two terms. The first term $E_{Y/X}\{C_b\min((y-(Q_1-x)^+)^+,I-(Q_1-x)^+)\}$ represents the expected costs of backorder during the replenishment lead-time. Since backorders are accepted only if they can be filled from replenishment, it is important to recognize that backorders during the replenishment lead-time can never exceed the effective inventory position after the first period backlog is cleared (i.e., $I-(Q_1-x)^+$). This condition is enforced by the operator $\min((y-(Q_1-x)^+)^+,I-(Q_1-x)^+)$. The second term of $C_2(I,x)$ is $E_{Y/X}E_{W/X}\{C_u(y+w-I)^+ +C_o(I-y-w)^+\}$. This represents the expected overstock and understock costs in the periods after the reorder is placed, until the end of the season.

It is important to recognize that $C_2(I, x)$ is neither convex nor concave in I. We illustrate this property using the following example.

Example 1: Let $Q_1 = 50$ and $x = 10$. Let the conditional probability distribution for $Y/(X = 10)$ be $P(Y/(X = 10) = 100) = 0.5$ and $P(Y/(X = 10) = 200) = 0.5$, while the conditional probability distribution for $W/(X = 10)$ be $P(W/(X = 10) = 100) = 0.5$ and $P(W/(X = 10) = 200) = 0.5$.
Let $C_b = 15$, $C_o = 20$, $C_u = 40$, $\lambda = 0.9$, $I^1 = 80$ and $I^2 = 110$, $I^\lambda = \lambda I^1 + (1-\lambda)I^2 = 83$. By substituting these values, using the values of Q_1 and x, and distributions Y/X and W/X to calculate expectations, it is easy to verify that:

$$C_2(I^\lambda,x) = E_{Y/X}[C_b\min((y-(Q_1-x)^+)^+,I^\lambda-(Q_1-x)^+)]$$
$$+E_{Y/X}E_{W/X}[C_u(y+w-I^\lambda)^+ +C_o(I^\lambda-y-w)^+]$$
$$= 9325 > \lambda C_2(I^1,x)+(1-\lambda)C_2(I^2,x) = 9317.5$$

This shows that $C_2(I, x)$ is not convex in I.

Next, let $I^1 = 80$ and $I^2 = 300$, so that $I^\lambda = \lambda I^1 + (1-\lambda)I^2 = 107$, while all the other values remain unchanged. Now,

$$C_2(I^\lambda,x) = E_{Y/X}[C_b\min((y-(Q_1-x)^+)^+,I^\lambda-(Q_1-x)^+)]$$
$$+E_{Y/X}E_{W/X}[C_u(y+w-I^\lambda)^+ +C_o(I^\lambda-y-w)^+] = 8672.5$$
$$< \lambda C_2(I^1,x)+(1-\lambda)C_2(I^2,x) = 8775.$$

This shows that $C_2(I, x)$ is not concave in I.

In view of this characteristic of $C_2(I, x)$, simulation based optimization techniques (Ermoliv and Wetts, 1998) typically used to compute the solution to these class of problems are not guaranteed to solve problem P2 and subsequently problem P1 to optimality. Consequently, for this reason and run time considerations, we elected to develop a heuristic, to address this problem. This heuristic is described in Sect. 3.4.

Once we have developed a scheme to solve this problem, to find the optimal reorder time, we would perform a line search on t to solve (P):

$$Z^* = \underset{0 \le t \le T-L}{\text{Min}} Z(t) \tag{P}$$

3.4   The Two Period Newsvendor Heuristic

The purpose of this heuristic is to set Q_1. In this regard, it is useful to understand the costs affected by the choice of Q_1. Firstly, a portion of Q_1 may remain unsold at the end of the season, generating an overstock. Secondly, during the interval 0 to $t + L$, if Q_1 is too small, we may incur backorder costs. During the interval t to $t + L$, we may also incur stockouts if satisfied and backordered demand exceeds $Q_1 + Q_2$, but it seems more natural to think of this cost as resulting from the choice of Q_2, not the choice of Q_1. Given this, we let $S = X + Y$, $U = X + Y + W$ and choose Q_1 to solve:

$$Z_h(t) = \underset{Q_1 \ge 0}{\text{Min}}\, \overline{C}(Q_1) = E_S C_b (s - Q_1)^+ \;+\; E_U C_o (Q_1 - u)^+ \tag{PH}$$

To solve this problem, let $F_1(s)$ and $F_2(u)$ be the distribution functions of random variables S and U respectively. The first order condition for problem (PH) is:

$$\frac{\delta Z_h(t)}{\delta Q_1} = -C_b\,(1 - F_1(Q_1)) + C_o F_2(Q_1) = 0$$

We set the heuristic order quantity Q_1^h to the value of Q_1 that satisfies this condition. Rearranging terms, this is calculated as the solution to the following equation:

$$F_1(Q_1) + \frac{C_o}{C_b} F_2(Q_1) = 1$$

Let $f_1(s)$ and $f_2(u)$ be the density functions of random variables S and U, respectively. Since $\dfrac{\delta^2 Z_h(t)}{\delta Q_1^2} = + C_b\, f_1(Q_1) + C_o f_2(Q_1) \geq 0$, the first order conditions are sufficient to establish the optimality of $Z_h(t)$ at Q_1^h. Note that our choice of Q_1^h minimizes expected backordering costs during the period before replenishment and minimizes expected overstock cost at the end of the season due to Q_1^h. The following result establishes conditions under which this heuristic finds an optimal solution.

Proposition 1: Suppose $Z(t)$ is the optimal solution to problem (P1) when random variables X, Y and W are perfectly correlated. Then, $Z_h(t) = Z(t)$.

Proof: If random variables X, Y and W are perfectly correlated, then $Y = \alpha X$ and $W = \beta X$, where α, β are positive constants. Thus, $E(Y/X = x) = \alpha x$, $V(Y/X = x) = 0$, $E(W/X = x) = \beta x$, and $V(W/X = x) = 0$. When all customers backorder, the optimal reorder quantity is $Q_2^* = [x(1+\alpha+\beta) - Q_1]^+$. If $Q_2^* > 0$, then we incur no overstock and understock costs in the third period after the reorder arrives. The only costs incurred will be possible backorder costs during the first two periods represented by $C_b[x(1+\alpha) - Q_1]^+$. If $Q_2^* = 0$, in addition to the backorder costs in the first two periods, we could incur an overstock of $[Q_1 - x(1+\alpha+\beta)]^+$ due to the initial order with associated costs $C_o[Q_1 - x(1+\alpha+\beta)]^+$. Consequently, total expected costs in the season when we have perfectly correlated demand can be expressed as $C(Q_1) = E_X\{C_b[x(1+\alpha) - Q_1]^+ + C_o[Q_1 - x(1+\alpha+\beta)]^+\}$. Since by definition, $S = X + Y = X(1+\alpha)$ and $U = X + Y + W = X(1+\alpha+\beta)$, $C(Q_1) = E_S\{C_b(s - Q_1)^+\} + E_U\{C_o(Q_1 - u)^+\} = \overline{C}(Q_1)$. Thus, $Z(t) = \underset{Q_1 \geq 0}{\text{Min}}\, C(Q_1) = \overline{C}(Q_1) = Z_h(t)$. Q.E.D.

It can be shown that Proposition 1 also holds under the assumption of no customer backorders. In light of this proposition, it is reasonable to expect good performance from this heuristic since the logical basis of implementing replenishment based on early sales is that demand during the later season is highly correlated with early demand. In our application, we found across all the products, the correlation between X and Y to be around 0.96 and between X and W to be about 0.95. This suggests that this heuristic could provide a simple and efficient basis to model the required decisions in this application.

Once we use the two period newsvendor heuristic to determine Q_1 and observe demand x during the first period, we approximate the optimal solution to the minimization problem P2 by setting $I = \max(I^*, Q_1-x)$ where I^* is the newsvendor quantity defined on $H_{R/X}$, the cumulative distribution of R updated by $X = x$ (i.e., $I^* = H_{R/X}^{-1}(\dfrac{C_u - C_b}{C_u - C_b + C_o})$). The quality of this approximation is also assessed in the application while evaluating the performance of the heuristic.

3.5   Modifications to Account for Returns of Merchandise

In catalog retailing, since customers place orders based upon photographs displayed in catalogs, purchased merchandise is often returned if the actual product differs from what the customer expected from the catalog. In this section, we describe how to extend our model to include merchandise returns.

Returned items can be resold if they are received before the season ends. This means that backorders in the interval $(0, t + L)$ and stockouts during the interval $(t + L, T)$ may be reduced by the availability of returns, but returns that are received too late to be resold can contribute to overstock. Based on the practice followed by the catalog retailer described in the application, we assume that a known fraction ω of customers return products, where $0 \leq \omega \leq 1$. These returns are immediately reusable if necessary to satisfy either a backorder or demand. We also assume that recycled returns (i.e. returns on returns and so on) are not reusable during the sales season. These assumptions ensure that we make an unbiased comparison with existing practice at this retailer.

Consequently, to adapt the heuristic to include returns of merchandise, we first consider the period until the reorder arrives. If Q_1 is the initial order quantity and s is the demand during this period, the total number of reusable returns is $\omega\min(s, Q_1)$. If $s > Q_1$, the total backorders that occur during this period are $[s - (Q_1 + \omega\min(s, Q_1))]^+ = [s - Q_1(1 + \omega)]^+$. Similarly, if u is the demand during the entire season, total reusable returns due to the initial order quantity are $\omega\min(u, Q_1)$. If $Q_1 > u$, the total overstock that occurs during the entire season is $[Q_1 - (u - \omega\min(u, Q_1))]^+ = [Q_1 - u(1-\omega)]^+$. Using these results, we redefine (PH) to:

$$Z_r(t) = \min_{Q_1 \geq 0} C_r(Q_1) = E_s C_b(s - Q_1(1 + \omega))^+ + E_U C_o(Q_1 - u(1 - \omega))^+ \quad (PH_r)$$

We use the procedure outlined in Sect. 3.4 to determine Q_1^h as the solution to the following equation:

$$F_1(Q_1(1+\omega)) + \frac{C_o}{(1+\omega)C_b} F_2(\frac{Q_1}{(1-\omega)}) = 1$$

At the end of the first period, we observe realized demand x and use it to set the reorder quantity $Q_2 = I - (Q_1^h - x)^+ - \omega\min(Q_1^h, x)$, where $I = \max(I^*, (Q_1^h - x)^+ + \omega\min(Q_1^h, x))$ and I^* is the newsvendor quantity defined on $H_{R'/X}$, the cumulative distribution of $R' = R(1-\omega)$ updated by $X = x$ (i.e., $I^* = H_{R'/X}^{-1}(\frac{C_u - C_b}{C_u - C_b + C_o})$).

3.6 Application

We have tested the ideas presented in this section at a large catalog retailer. We applied the model and the two-period newsvendor heuristic to make purchase decisions for 120 style/colors from the women's dresses department appearing in a particular catalog. We chose this division because it represented a significant portion of the business. The sales season for these products is T = 22 weeks and the replenishment lead-time is L = 12 weeks. As these products are sold through mail order catalogs, the price during the season is fixed. Around 35% of sales are returned, i.e. $\omega = 0.35$.

In the process currently in place at this retailer, initial order quantities are set to forecast demand for the 22-week season adjusted for anticipated merchandise returns. Forecasts for each style/color are updated after 2 weeks by dividing observed sales by the historical fraction of total season sales for this department that have been observed in the past to normally occur in the first 2 weeks. Reorders are placed to make up the difference from an updated forecast adjusted for returns. Specifically for a given style/color, if f is the total forecast sales and ω is the anticipated fraction of returns, then $Q_1 = (1-\omega) f$. Letting x_2 be the actual sales observed at the end of 2 weeks and k_2 be the fraction of total demand historically observed at this point for a group of similar products, then we set $Q_2 = ((1 - \omega)x_2/k_2 - Q_1)^+$. Note that this procedure sets Q_1 to the forecast sales net of anticipated returns during the entire season and, hence, reorders are used as a reaction to larger than anticipated sales rather than something that is planned for in advance.

Figure 2 shows the improvement in forecast accuracy due to updating at this retailer. Each point shows forecast and actual demand for a particular style/color combination. The left graph compares demand forecasts with actual demand, for the average of forecasts made by four expert buyers prior to the beginning of the season. In the right graph, the forecasts equal actual sales after 2 weeks into the season divided by a factor representing the fraction of total sales historically observed after 2 weeks.

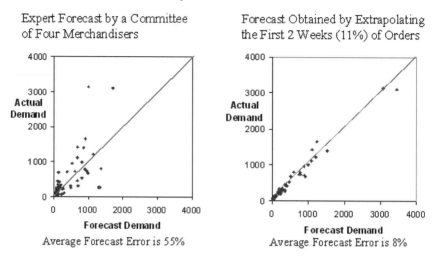

Figure 2. Comparison of early and updated forecasts

Application of our model requires a method to estimate demand probability distributions. This is particularly challenging as there was no sales history for any of the new dresses. However, we were able to calculate forecast errors defined as the difference between buyer forecast and actual sales for similar products appearing in the same catalog from the past 2 years. We used this information to conclude that the distribution of forecast errors were normally distributed with a large degree of confidence (χ^2 test holds at $\alpha = 0.01$ level). We assumed that forecast errors would follow a similar distribution in past and future seasons. This seemed reasonable as the same individuals who forecast product demand in the past seasons were also forecasting current season demand.

Since the demand for any given product is equal to its forecast plus the associated forecast error, this implies that the demand distribution for U for a given product during the entire season is normally distributed. While probability distributions for retail products seem to have long tails, these result from plotting *actual* demand for products that seem indistinguishable (or at least similar) ex ante. However, in contrast, U represents the demand distribution for a given product.

To estimate normal parameters μ and σ of this distribution, we implemented the procedure developed by Fisher and Raman (1996). In this method, the members of a committee (comprised of four buyers in our case), independently provide a forecast of sales for each product. The mean μ is set to the average of these forecasts. The standard deviation of demand σ is set to $\theta\sigma_c$, where σ_c is the standard deviation of the individual committee member forecasts and the factor θ is chosen so that the average standard deviation of historical forecast errors equals the average standard deviation assigned to new products. In our application, we found θ to be 1.4.

To estimate the parameters of distribution X and R, where $U = X + R$, we assume that (X, R) follows a non-degenerate bivariate normal distribution. For this distribution, it is well known (Bickel and Doksum, 1977) that the marginal distribution of X is an univariate normal distribution with mean μ_x and standard deviation σ_x, while the marginal distribution of R is also normally distributed with mean μ_R and standard deviation σ_R. Let k_t represent the proportion of total sales until reorder point t, δ_t the correlation between X and R and ρ_t the correlation between X and U. We estimate k_t, δ_t and ρ_t from historical data and use the formulas developed in Fisher and Raman (1996) to calculate $\mu_x = k_t\mu$, $\sigma_X = \sigma[\rho_t - \delta_t\sqrt{\dfrac{(1-\rho_t^2)}{(1-\delta_t^2)}}]$,

$\mu_R = k_t(1-\mu)$ and $\sigma_R = \sigma\sqrt{\dfrac{(1-\rho_t^2)}{(1-\delta_t^2)}}$.

For the bivariate normal distribution (X,R), note that the updated distribution R/X = x is also normally distributed with mean $\mu_{R/X} = \mu_R + \delta_t$ $(x - \mu_x)\sigma_R/\sigma_x$ and standard deviation $\sigma_{R/X} = \sigma_R\sqrt{1-\delta_t^2}$. Since $0 \le \delta_t < 1$, this implies that $\sigma_{R/X} \le \sigma_R$. Thus forecast updating based on actual sales x reduces variance in the distribution of demand during the remaining season and permits a more accurate forecast. By using replenishment, the retailer can take advantage of this improved forecast by placing a more precise reorder that directly contributes to higher expected profits during the remaining season.

To better understand the nature of forecast errors, we compared the standard deviation of the committee forecast for individual products at the beginning of the season (i.e. σ_c) with its corresponding forecast error. These results, shown in Fig. 3 suggest that when the committee agrees they tend to be accurate and that the committee process is a useful way to determine what you can and what you cannot predict.

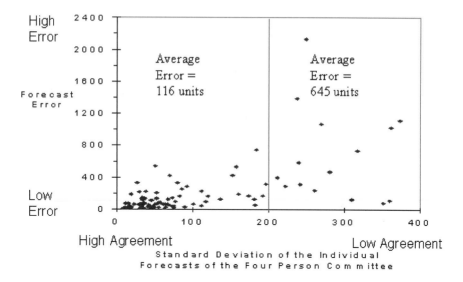

Figure 3. Committee standard deviation vs. forecast error

With the exception of the backorder penalty C_b, all the cost parameters required for our analysis were readily available. Estimating the backorder penalty is challenging in practice because in addition to the $1 per unit extra transaction cost for procurement and distribution associated with a backorder, there is an intangible cost due to customer ill will. The company was uncertain as to the exact value of the ill will cost, but felt a value of C_b in the range $5 to $15 was reasonable. We applied our analysis to three cases using $5, $10 and $15 per unit as values of C_b. We also analyzed the case $C_b = 1$ to insure that our heuristic did not outperform the current rules because we assessed an ill will cost which was not used in the current rule.

Note that although ill will costs can also be added to C_u, we did not, as an ill will cost in this application was charged only because management was mainly interested in insuring the flexibility to backorder was not abused. Given the values of C_u and C_o, if $C_b = 0$, then it is optimal to set $Q_1 = 0$ and backorder all first period demand. But these excessive backorders would likely reduce market share in the long term. The omission of ill will costs in C_u does not affect the analysis, since, depending on the product, C_u was 2–4 times greater than backorder costs, and consequently, it would never be optimal not to satisfy demand to avoid a backorder.

As a practical matter, we found at this retailer that historically around 5% of customers chose not to accept the offer to backorder. Consequently, we adjusted the backorder cost to account for this fraction of lost sales by defining an effective backorder cost $C'_b = 0.95*C_b + 0.05*C_u$, representing the costs of a backorder and stockout weighted by the expected fraction of customers who would choose either option. We replaced C_b with C'_b in the definition of problem (PH_r).

The first step in our methodology is to determine the reorder time. It is important to accurately choose this time since if chosen too early in the season, actual sales are not sufficient to provide an accurate revision of the second period demand forecast. On the other hand, if the reorder time is too far into the season, the benefit of replenishment is diminished as it is now used to service only a small portion of the season. The specific choice of reorder point depends on the proportion of total sales observed during the initial weeks and the length of the replenishment lead-time. For instance, if this proportion is high and/or there is a long lead-time, one would choose the reorder point early on the season to ensure that a reasonable proportion of total sales is serviced by the reorder.

To determine the best reorder time, we used Monte Carlo simulation with the estimated distributions of demand to calculate $Z(t)$ for week t. In this procedure, for a given reorder time, we estimate the initial order quantity using our heuristic. We simulate x, as a realization of X, the distribution of total demand until the reorder placed. We use x to calculate the backorder costs before the reorder is placed, update $R/X = x$ and calculate the expected costs during the remaining season. We repeat this procedure for several simulated realizations of X and calculate the expected costs during the entire season associated with a reorder time by averaging the costs associated with each realization. As discussed previously, using the bivariate normal distribution to model demand (X, R) ensures that both X and R are univariate normal distributions and the variance of updated second period demand $R/X = x$ is also univariate normal whose variance is now reduced from σ_R to $\sigma_{R/X} = \sigma_R \sqrt{1 - \delta_t^2}$.

We repeat the simulation for several choices of reorder time. The results of this simulation are summarized in Fig. 4. Since $Z(t)$ attains its minimum at $t = t* = 2$, the reorder time is chosen to be at the end of week 2. Note that the length of the replenishment lead-time assumed in this analysis is 12 weeks. As the season lasts only 22 weeks, we cannot reorder after week 10. Consequently, the value of $Z(t)$ for $t \geq 10$ is set equal to the expected costs incurred for a single period buy if we set Q_1 to Q_1^s the newsvendor quantity defined on the total distribution of demand (i.e., $Q_1^s = F_2^{-1}(\dfrac{C_u}{C_u + Co})$).

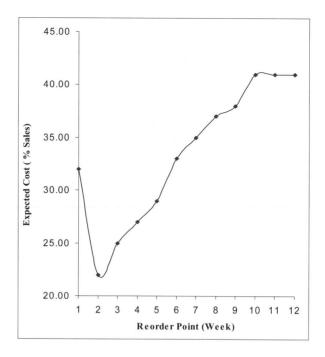

Figure 4. Reorder time and expected costs

A key factor that influences the level of profits gained by replenishment is the proportion of total demand over time observed during the early part of the season and the time until the reorder arrives. Clearly, if this proportion is very high, then the benefits of replenishment are limited, as the reorder would only a serve a small proportion of total season demand. In our application, we found that historically for similar product lines 10% of total demand is observed when the reorder is placed after 2 weeks and 50% of demand is observed at week 10 when the reorder arrives. These values confirmed that replenishment based on actual sales was a viable strategy for the chosen product line and motivated us to apply our method to determine initial and replenishment order quantities.

For the 120 style/colors in this department, we determined that the initial order quantity by solving problem (PH_r) using the heuristic modified to include returns. We then observed x, the sales until the second week and set the reorder quantity Q_2 using the procedure developed in Sect. 3. Since at the end of the season we knew total sales and actual sales per week for each dress, we were able to calculate the stockouts, overstock, backorders, dollar sales and profits that would have resulted from our ordering policy. To compare our method with the current ordering rules, we also calculated these values for the current policy. The results consolidated across all the 120 dresses are tabulated in Table 5.

Observe that although the total orders placed by our method and existing practice are similar, the composition of these orders across the two periods is different. Our heuristic reduces overstocks, stockouts and backorders enough to increase profits compared with the current rule by from 2.23 to 4.92% of current sales, depending on the value of C_b. Profit before tax for this retailer is around 3% of sales. Consequently, our heuristic offers the potential to approximately double profit.

Table 5. Comparison of the two period newsvendor heuristic with current practice

	Current rule	Model CiB = $5	Model CiB = $10	Model CiB = $15
Initial order	19,050	14,479	18,015	20,680
Reorder	2,816	5,229	3,931	3,179
Total buy	21,866	19,708	21,946	23,859
Overstock	7,689	4,924	7,061	8,925
Omits	3,712	2,998	2,566	2,423
Backorders	6,534	8,643	6,969	5,989
Profit ($)	431,696	496,597	445,384	395,782
Sales ($)	1,317,889	1,301,579	1,372,926	1,422,108
Profit increase by model (as % current sales)		4.92	3.52	2.23

Our results also show the impact of C_b on the solution. Order quantities for the current rule do not change, as the current rule does not consider C_b in determining order quantities. When $C_b = \$5$, we order a smaller initial quantity as backorders are now relatively less expensive. This in turn increases backorders and stockouts, but reduces overstocks, which increases profit improvement from 3.52 to 4.92%. On the other hand when $C_b = \$15$, we order a larger initial quantity as backorders are relatively more expensive. This reduces backorders and stockouts, but increases overstocks, which reduces the profit improvement from 3.52 to 2.23%.

To insure that our heuristic did not outperform the current rules because we assessed an ill will cost which was not used in the current rules (possibly because ill will costs were not recorded in the books), we also considered the case when $C_b = 1$. Here, C_b only consists of the additional transaction cost per unit of procurement and distribution associated with a backorder. As expected, our heuristic ordered a substantially smaller amount initially than the cases with larger values of C_b. This in turn increased backorders and stockouts, but reduced overstocks. Over all, the profit improvement over the current rules increased to 5.52%. This is consistent with the general pattern in Table 1 that shows profit improvement increasing as C_b decreases.

This analysis assumes a replenishment lead-time of 12 weeks. It is easy to understand that reducing this time could potentially increase the benefits of replenishment, as a greater portion of the season can be serviced from the more accurate reorder. However, it is important to precisely calculate this benefit to justify the costs of lead-time reduction. Our methodology provides a framework to analyze these benefits both before and after sales are realized. To perform this analysis before actual sales are realized, we use a simulation to calculate expected costs for different values of lead-times. Using actual sales and for the case in which $C_b = 10$, we performed an analysis identical to the one used to obtain the results reported in Table 1, but using the new choices of the lead-time. These results are summarized in Fig. 5 and indicate that the length of the replenishment lead-times significantly influences the benefits of replenishments. This type of analysis

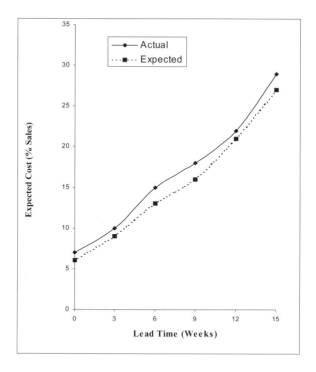

Figure 5. Replenishment costs and lead times

could be used to decide between a domestic supplier with typically higher costs but shorter lead-times, and a foreign supplier with relative low costs but long lead-times.

To further evaluate the quality of this heuristic, we solved (P1) using a simulation-based optimization method [26]. In this technique, we use simulation to numerically compute $C(Q_1)$ for selected values of Q_1 in the range $[0, Q^*]$, where $Q^* = F_2^{-1}(\dfrac{C_u}{C_u + C_o})$ is the newsvendor quantity [27] defined across the whole distribution of demand. Finally, we set $Z(t) = \underset{0 \leq Q \leq Q^*_1}{\text{Min}} C(Q_1)$. In the case where $C_b = 10$, this technique improved profit relative to the current rule by around 3% of current sales, a lower improvement than was achieved with our heuristic. In the cases where C_b equaled 5 or 15, the profit improvement was also marginally less than achieved with our heuristic.

In addition to resulting in a smaller profit gain than the two period newsvendor heuristic, we found that the solution time for this technique was around 40 h on a Dell Pentium II PC as compared with less than a minute for the two period newsvendor heuristic. These results provide strong justification for using this heuristic in this application.

We also considered the impact on performance of correlation between early and later demand. In our application, across all products, we found the correlation between X and Y to be around 0.96 and between X and W to be about 0.95. In view of Proposition 1, such high correlation suggests that the heuristic solution is very close to the optimal solution to this problem.

We performed a computational study to evaluate the performance of the heuristic for different levels of correlation between early and later season demand. Define ρ_1 as the correlation between X and Y, and ρ_2 as the correlation between X and W. For simplicity, we set $\rho_1 = \rho_2 = \rho$ and vary ρ from 0 to 0.99 in steps of 0.1. For a given value of ρ, and using the committee estimates of μ and σ for each product, we calculated Q_1 using the heuristic and using the simulation-based optimization method. We then used these values of the initial order quantities and Monte Carlo simulation with

---

[26] Please refer to Ermoliv, Y. and R.J.B. Wetts (1998), "Numerical Techniques for Stochastic Optimization" Springler Verlag, New York, for a theoretical justification and a detailed description of this technique. The same simulation based technique is used to numerically estimate $C_2(Q_1, x)$ required in the computation of $C(Q_1)$. Here we vary I over the range $[0, 10Q^*]$ and set $C_2(Q_1, x) = \underset{0 \leq I \leq 10Q^*}{\text{Min}} C_2(I)$.

[27] We choose this value as it is highly unlikely that an initial order quantity greater than this quantity would not be sufficient to cover sales during the periods before the replenishment arrives.

the estimated distribution of U for each product to calculate expected costs for each technique.

For a given value of ρ, let C_h represent the total expected cost of the heuristic across all products, and C_s the corresponding total cost of the simulation based optimization procedure. The percentage performance gap of the heuristic is defined as $(C_h - C_s)/C_s \times 100\%$. We report the percentage performance gap across a range of values for demand correlation in Table 6. For each level of demand correlation, the run times for the heuristic across all the products was less than a minute, while the equivalent run time for the optimization approach was around 40 h.

The results in Table 6 show that this gap varies from 0.06 to 5.2%, with the highest gaps occurring at the lowest levels of correlation. These results suggest that the heuristic provides an efficient basis to address this problem, even for cases that have modest levels of correlation (e.g. $|\rho| > 0.3$) between early and later sales. Since fashion replenishment makes most sense for products with some degree of correlation between early and later sales, this heuristic seems to be an accurate, simple and intuitive way for a retailer to implement this strategy.

Table 6. Percentage performance gap between the two period newsvendor heuristic and a simulation based optimization method for different levels of demand correlation

Correlation	Performance gap (%)	Correlation	Performance gap (%)
0	5.2	0	5.2
−0.1	4.15	0.1	4
−0.2	3.1	0.2	3
−0.3	1.8	0.3	1.7
−0.4	1.5	0.4	1.4
−0.5	1.2	0.5	1.3
−0.6	1.1	0.6	0.9
−0.7	0.8	0.7	0.7
−0.8	0.5	0.8	0.6
−0.9	0.3	0.9	0.2
−0.99	0.1	0.99	0.07

4.   Conclusions

Our goal in this chapter has been to expose the reader to an intellectually interesting problem context laden with opportunities for research that can have a high impact on retailer profits. Following are some conclusions and future research directions that can be drawn from this chapter.

In the area of merchandise testing, we find that the sales of a given product mix vary greatly among the stores of a large chain. Some, but not all, of this variation can be explained by store descriptors such as average temperature and ethnicity. A merchandise testing process that exploits this by clustering stores based on similarity of sales mix and choosing a single store within each cluster can provide forecasts of season demand for style/colors accurate to about 10–20%. This approach performs significantly better than alternative methods used in retail practice, based on standard statistical approaches or on clustering by store descriptor variables. The impact on cost of the superior performance is enough to double a retailer's profits. The merchandise testing problem invite additional research on the following topics. First, choosing a small number of test stores from the many stores of a large chain is analogous to the statistical problem of choosing a parsimonious set of independent variables from a large potential set in a regression. The ability of the k-median approach to find variables that are minimally co-linear, and hence more predictive, deserves study in this broader context. Second, the relationship between the clusters formed by our algorithm and micro-merchandising is worth exploring. Each cluster could be treated as a "virtual chain" within the larger chain, which is managed separately and in a consistent manner in terms of product mix, timing of delivery, advertising message, store layout, etc. While this may require more managerial effort and careful attention to detail, the increasingly sophisticated information systems being installed by many retailers offers the opportunity to gather the required information and automate much of the tedious data analysis tasks.

In the context of optimizing replenishment of retail fashion products, we find that an ounce of actual sales data is worth more than a pound of estimation in improving demand forecasts. Replenishment based planning provides a mechanism to use actual sales data during the season. In our application at a catalog retailer, we found that compared to no replenishment, a single optimized replenishment improves profit by a factor of five. However, the benefits of replenishment are significantly influenced by the choice of the demand update point and the length of the replenishment lead time. The replenishment planning problem for fashion products invites additional research in the following topics. First the approach presented here could be extended to include multiple products with capacity and constraints and order minimums. These extensions become critical when this approach is applied to categories of branded merchandise that are typically made by a single manufacturer and sold at several competing retailers. Second, our approach could be modified to include variable pricing decisions during the sales season. This aspect is important when this model is applied to store format retailers.

In conclusion, we believe that the models described in this chapter provides a powerful framework to improve the accuracy and analyze several crucial aspects of the merchandise planning process for retail fashion products.

References

Anthony, R. and Loveman, G., Laura Ashley and Federal Express Strategic Alliance. *Harvard Business School Case* N9-693-050 1994.

Aufreiter, N., Karch, N., and Smith Shi, C., The Engine of Success in Retailing. *The McKinsey Quarterly*, 3, 1993, 101–116.

Bickel, P. and Doksum, K., 1977, *Mathematical Statistics*, Holden Day Publishers, San Francisco.

Bitran, Gabriel R., Elizabeth Haas and Hirofumi Matsuo, 1986, Production Planning of Style Goods with High Setup Costs and Forecast Revisions, *Operations Research*, 342, 221–246.

Bradford, John W. and Paul K. Sugrue, 1990, A Bayesian Approach to the Two-period Style-goods Inventory Problem with Single Replenishment and Heterogeneous Poisson Demands, *Journal of Operations Research Society*, 413, 211–218.

Brooke, A., Kendrick D., and Meeraus, A., *GAMS: A User's Guide*, The Scientific Press, San Francisco, 1992.

Cornuejols, G., Marshall L. Fisher, and George L. Nemhauser, Location of Bank Accounts to Optimize Float: An Analytic Study of Exact and Approximate Algorithms, *Management Science*, 23, 1977, 789–810.

DiRomualdo, R., CEO, The Borders Group, The Borders System, talk given at the Borders Annual Meeting, 1998.

Doyle, P. and Gidengil, B.Z., A Review of In-Store Experiments, *Journal of Retailing*, 532, 1977, 47–62.

Economist: A Survey of Retailing, Change at the Check-Out. *Economist* March 4th 1995.

Eppen, Gary D. and Ananth V. Iyer, 1997a Improved Fashion Buying using Bayesian Updates, *Operations Research*, 456, November–December.

Eppen, Gary D. and Ananth V. Iyer 1997b, Backup Agreements in Fashion Buying – The Value of Upstream Flexibility, *Management Science*, 4311, 1469–1484.

Ermoliv, Y. and Wetts, R.J.B., 1998, *Numerical Techniques for Stochastic Optimization*, Springer Verlag, New York, NY.

Fisher, M. and Raman, A., 1996, Reducing the Cost of Demand Uncertainty through Accurate Response to Early Sales, *Operations Research*, 444, 87–99.

Fisher, M.L., Raman, A., and McClelland, A.S., 2000, Rocket Science Retailing, Working Paper, *Harvard Business Review*, 78, 115–124.

Fisher, M.L., and Rajaram, K., 2000, Accurate Retail Testing of Fashion Merchandise: Methodology and Application, Working Paper, The Anderson School at UCLA.

Fisher, M. and Rajaram, K., Summer 2000, Accurate Retail Testing of Fashion Merchandise: Methodology and Application, *Marketing Science*, 19, 3, pp. 266–278.

Fisher, M.L., Rajaram, K., and Raman, A., Optimizing Inventory Replenishment of Retail Fashion Products, *Manufacturing and Service Operations Management*, 3, 3, pp. 230–241, Summer 2001.

Fox, L.J., November 1995, An Integrated View of Assortment Management Process: The Next Frontier for Leading Retailers. *Chain Store Age Executive*.

Frazier, R.M., January, 1986, Quick Response in Soft Lines, *Discount Merchandiser*, 42–56.

Hammond, J.H., 1992, Dore'-Dore'. *Harvard Business School Case* N9-892-009.

Hammond, J.H., 1990, Quick Response in the Apparel Industry. *Harvard Business School Note* N9-690-038.

Hausman, Warren H. and Rein Peterson, 1972, Multiproduct Production Scheduling for Style Goods with Limited Capacity, Forecast Revisions and Terminal Delivery *Management Science*, 187, 370–383.

Hollander, Stanley C., 1986, A Rearview Mirror Might Help Us Drive Forward, *Journal of Retailing*, 621, 7–10.

Hooper, L. and Yamada K., September 14, 1992, IBM Signals Delay Affecting Low-Cost Line. *Wall Street Journal*.

Matsuo, H., 1990, A Stochastic Sequencing Problem for Style Goods with Forecast Revisions and Hierarchical Structure, *Management Science*, 363, 332–347.

Murray, George R. Jr., and Edward A. Silver, 1966, A Bayesian Analysis of the Style Goods Inventory Problem, *Management Science*, 11, 785–797.

Myers, R.H., 1990, *Classical and Modern Regression with Applications*, PWS-KENT Publishing Company, Boston.

Pashigan, P.B., Demand Uncertainty and Sales: A Study of Fashion and Markdown Pricing. *American Economic Review*, Vol. 78, 1988, 936–953.

Patterson, Gregory A., Target Micromarkets its way to success: No 2 Stores Are Alike, *Wall Street Journal*, May 31, 1995.

Pollack, Elaine, Raising the Bar: Keys to High Performance, *Chain Store Executive Age*, 1994.

Raman, A, 1999, Managing Inventory for Fashion products, *Quantitative Models for Supply Chain Management*, Kluwer Academic Publishers, Chapter 25.

San Juan Star U.S. Retailers Slash Prices to Lure Shoppers, December 27, 1996.

Stalk, G., Evans, P., and Shulman, L.E., March–April 1992, Competing on Capabilities: The New Rules of Corporate Strategy. *Harvard Business Review*, 57–69.

Standards and Poors Industry Survey: Retailing Specialty Apparel, January 22, 1998.

Stewart, T.A., September 21, 1992, Brace for Japan's Hot New Strategy. *Fortune*.

Urban, G.L., and Hauser, J.R., *Design and Marketing of New Products*, Prentice-Hall Inc., New Jersey, 1980.

USA Today. What you may not get what you want for Christmas, December 24, 1996.

White, J.B., August 7, 1992, Hughes Charge Gives GM Net Loss of $357.1 Million, *Wall Street Journal*.

Wilson, B.L., Kingdom, M., and Reeve, T., November 1995, Quick Hits in Store Level Merchandise and Inventory Management, *Chain Store Age Executive*, 103–106.

Supply Chain Management in the Presence of Secondary Market

Hau L. Lee, Barchi Peleg, Seungjin Whang, and Yan Zou
Graduate School of Business, Stanford University, Stanford, CA 94305-5015, USA

Abstract: This paper studies the impacts of online secondary markets on a supply chain. Retailers who recognize that there is a secondary market to dispose of their excess inventory, might make higher purchase quantities in the beginning of a product life cycle. On the other hand, given that there is a secondary market that one can tap into for future needs, a retailer may reduce its initial purchase quantities. We endogenously derive the optimal decisions for the retailers, and explore the impacts of the secondary market on the performance of the supply chain as a whole. Next, we explore the impact of a manufacturer's participation in the secondary markets, together with the retailers, for a special kind of secondary market known as diversion. In addition, we study the role of the secondary market as a mechanism to capture demand information through the realized price. Lastly, we discuss managerial implications of secondary markets and offer further research directions.

Key words: Supply chain management; electronic markets; inventory.

1.   Introduction

E-marketplaces, Internet destinations that bring buyers and sellers together in a web based exchange to conduct electronic commerce, have been hyped as the future way of business-to-business (B2B) commerce. But in only a few years' time, the majority of the e-marketplaces have either ceased to exist, or have been merged with one other. One of the few types of e-marketplaces that have continued to flourish is the one used by companies/individuals to dispose of and acquire excess or used assets. We are familiar with the success of eBay that, among others, serves as a secondary market for individuals and enterprises. Other examples of secondary markets, used primarily for B2B commerce, include Covisint for the automotive industry; Converge, which focuses on the high-tech industry; and Exostar, which targets the aerospace and defense industry. Secondary markets are especially popular in industries such as high-tech, which are plagued with highly uncertain demands, continuously shrinking product life cycles and high risks of product obsolescence. The Internet facilitates the operation of secondary markets as it can easily bring thousands of buyers and sellers together to make transactions, at minimal search and transaction costs. Unlike primary purchases that are often conducted based on long term supplier relationships or special contract terms, transactions in secondary markets are often based on price and availability only, and are therefore much easier and simpler to implement.

Several lines of research are related to secondary markets. Auctions have been used extensively as the trade mechanism in secondary markets, and there has been a tradition of economics research on the theory of auctions (e.g., Milgrom and Weber, 1982). Research of auctions in e-marketplaces can be found, for example, in Gallien and Wein (2000). Our current paper is not about the mechanics of on-line secondary markets, but rather on the impact of such markets on operational performance of manufacturers and retailers. Therefore, more relevant lines of research include inventory pooling (e.g., Eppen and Schrage, 1981), or lateral transshipment in the multi-echelon literature (see Lee, 1987), and component commonality in the operations management area (e.g., Gerchak et al., 1988). The past research was, however, limited in their direct applicability to online secondary markets due to their focus on the optimization from one party's perspective. In addition, the number of retailers considered was limited, and a common assumption was that market price in the use of inventory pooling or commonality is exogenously given.

This paper has a dual objective: first, to give an overview of ongoing research related to the impact of secondary markets to a supply chain and second, to offer general discussions on secondary markets. In Sect. 2 we start with studying how secondary markets change the retailers' purchase decisions, as well as the impact on the consumers and the manufacturer whose goods the retailers sell. In Sect. 3 we examine the impacts of a manufacturer's participation in the secondary markets, together with the retailers, for a special kind of secondary market known as diversion. In Sect. 4 we explore the informational role of secondary markets where the market price serves as a signal of demand information to the retailers. In Sect. 5, we offer managerial discussions on secondary markets in the context of supply chain management. Due to space limitations, all proofs are omitted. They are available upon request from the authors of this paper. Details of the works in this paper can also be found in Lee and Whang (2002), Peleg and Lee (2002) and Whang and Zou (2003).

2. Impacts of Secondary Market in the Supply Chain

Consider n retailers who order a single product from the manufacturer and sell it over a short sales season (typical of the high-tech or fashion industry). At the beginning of the season, retailer i orders Q_i from the manufacturer at a unit price p_1, which is delivered immediately. Retailer sales z_i are realized over the sales season where z_i is a random variable independently drawn from $F(\cdot)$, a twice differentiable distribution function over $[0,\infty)$ with a finite mean μ and finite variance. Each unit is sold at π to end consumers, and the salvage value of the units left unsold is zero. Unfilled orders are lost. Under this standard setting for the newsvendor problem, the optimal stock level Q_i^0 for retailer i is:

$$Q_i^0 = F^{-1}\left(\frac{\pi - p_1}{\pi}\right) \tag{1}$$

Consider now a modified model where a secondary market opens at some point during the sales season, dividing it into *periods* 1 (prior to the opening of the secondary market) and 2 (subsequent to the opening of the secondary market). At the beginning of period 1, retailer i orders Q_i from the manufacturer, which is delivered immediately. First-period retailer sales x_i are realized. At the end of period 1 the secondary market is opened where retailers trade units at a uniform price p_2, which is endogenously determined to clear the market. Given p_2, each retailer chooses a stock level q_i for the second period. Sales y_i are realized in period 2. x_i and y_i are random variables independently drawn from distribution functions $F_1(\cdot)$ and $F_2(\cdot)$

with respective means μ_1 1 and μ_2. We assume that x_i and y_i are independent for each i. To be consistent with the model with no secondary market, F is the convolution of F_1 and F_2. We consider the case where n is infinitely large (which mimics an Internet-based market), so each retailer is a price-taker in the secondary market: i.e., $d\,p_2/d\,Q_i = 0$. The equilibrium solution is defined as follows:

Second-period stock level: Given any $\mathbf{Q}^{28}$, $\mathbf{x}$, and p_2, retailer i chooses q_i^* to maximize his expected profit for period 2:

$$q_i^* = \arg\max\left\{\pi\int_0^{q_i} yf_2(y)dy + \pi q_i\int_{q_i}^{\infty} f_2(y)dy - p_2\left(q_i - \left(Q_i - x_i\right)\right)\right\}$$

$$= F_2^{-1}\left(\frac{\pi - p_2}{\pi}\right)$$
(2)

Given identical retailers, $q_i^* := q^*$ for each i.

Equilibrium price in period 2: Given any $\mathbf{Q}$ and $\mathbf{x}$, total demand is nq^* whereas total supply is $\sum_{i=1}^{n}\left(Q_i - x_i\right)^+$. The market price p_2 is determined to have demand equal supply and clear the market. This, together with (2), yields p_2 satisfying:

$$p_2 = \pi\left[1 - F_2\left(\frac{1}{n}\sum_{i=1}^{n}(Q_i - x_i)^+\right)\right]$$
(3)

By symmetry, $Q_i = Q$ for all i. If n is infinitely large, then the law of large numbers yields:

$$\lim\frac{1}{n}\sum_{i=1}^{n}(Q_i - x_i)^+ = E(Q-x)^+ = \int_{x=0}^{Q}(Q-x)f_1(x)\,dx$$

$$= \int_0^Q F_1(x)dx := \Gamma_1(Q),$$

---

[28] A boldface letter denotes the n-vector of corresponding variables: e.g., $\mathbf{Q} = \left(Q_1, Q_2, \cdots, Q_n\right)$

where $\Gamma_i(Q) := \int_0^Q F_i(x)dx$ for $i = 1,2$ throughout the paper. From this and Slutsky Theorem, we obtain the limiting price $\hat{p}_2$ as follows:

$$\hat{p}_2 := \lim_{n \to \infty} p_2 = \pi[1 - F_2(\Gamma_1(Q))]. \tag{4}$$

Optimal order quantity in period 1: To derive the symmetric Nash equilibrium first period stock levels $\mathbf{Q}^\cdot$, we take other retailers' decisions $\mathbf{Q}^\cdot_{-i} := (Q^*, Q^*, ..., Q^*)$ as given and derive retailer i's optimal decision Q_i. Letting for short $p^\cdot_2(Q_i) := p_2^* (\mathbf{Q}^\cdot_{-i}, Q_i)$ and $q^\cdot(Q_i) := q^\cdot(\mathbf{Q}^\cdot_{-i}, Q_i)$, retailer i's decision problem is given by:

$$\max_{Q_i} \left\{ \pi \int_0^{Q_i} x f_1(x)dx + \pi Q_i \int_{Q_i}^\infty f_1(x)dx - p_1 Q_i + v_2(Q_2) \right\}$$

where

$$V_2(Q_i)$$
$$= E\left[\pi \int_0^{q^*(Q_i)} y f_2(y)dy + \pi q^*(Q_i) \int_{q^*(Q_i)}^\infty f_2(y)dy + p_2^*(Q_i)((Q_i - x_i)^+ - q^*(Q_i)) \right]$$

As n approaches infinity, $d\, p_2^*(Q_i)/dQ_i = dq_2^*(Q_i)/dQ_i = 0$, due to the price-taker assumption, and thus we can use the definition $p_2^* := p_2^*(Q_i)$ and $q^* := q^*(Q_i)$. Therefore,

$$\lim_{n \to \infty} V_2'(Q_i) = p_2^* \frac{d}{dQ_i} \int_0^{Q_i} (Q_i - x_i) f_1(x)dx = p_2^* F_1(Q_i).$$

Using this, and by differentiating the objective with respect to Q_i we obtain:

$$F_1(Q_i) = \frac{\pi - p_1}{\pi - p_2^*} \tag{5}$$

The symmetric equilibrium for period 1 is thus $\mathbf{Q}^* = (Q^*, Q^*, ..., Q^*)$, where Q^* satisfies (5).

We next study whether a secondary market will increase or decrease the sales of the manufacturer in the first period. It turns out the answer to this question is an outcome of a complex tradeoff among multiple factors. In short, the secondary market may increase or decrease the manufacturer's sales, depending on the critical fractile $\dfrac{\pi - p_1}{\pi}$.

THEOREM 1. *In equilibrium the secondary market increases the manufacturer's sales (i.e., $Q^0 \leq Q^*$) if and only if*

$$F_1(F^{-1}(\alpha)) \cdot F_2\left(\Gamma_1\left(F^{-1}(\alpha)\right)\right) \leq \alpha, \qquad (A)$$

where α is the critical fractile $\dfrac{\pi - p_1}{\pi}$.

THEOREM 2. *If $\lim_{x \to 0+} f(x) > 0$, there exists a strictly positive α' for which if $\alpha \leq \alpha'$, the secondary market increases the manufacturer's sales.*

Theorems 1 and 2 show that the secondary market has the effect of "regression towards the mean." That is, it dilutes the impact of the critical fractile on the retailer's decisions, since the retailer orders more than the newsvendor solution if the critical fractile is small, but less if the critical fractile is large. The phenomenon is attributed to the *price effect* of the secondary market. Consider a product with a small critical fractile. The one-shot newsvendor solution suggests a smaller order quantity by the retailers and a (stochastically) lower leftover inventory at the end of the first period. As a result, the price in the secondary market will likely be relatively high, and this will invite retailers to buy more initially from the manufacturer than the newsvendor solution. Symmetrically, for a high-fractile product, the one-shot newsvendor solution suggests a higher order quantity and a stochastically higher leftover inventory at the end of the first period. Thus, the price in the secondary market will likely be low, and this encourages retailers to purchase less initially from the manufacturer. If condition A is not met, each retailer will buy less than the newsvendor solution and the manufacturer will be worse off as a result of the secondary market.

We next study the impact of the secondary market on the performance of the supply chain as a whole. The results are summarized in Theorem 3.

THEOREM 3. *For given positive Q^0 and Q^*, the following holds:*

(a) There exists $\underline{\Delta}(<0)$, such that:

(i) The expected sales in the channel with a secondary market is larger than that without a secondary market if and only if
$$Q^* - Q^0 > \underline{\Delta}$$

(ii) The expected number of stockouts in the channel with a secondary market is smaller than that without a secondary market if and only if $Q^0 - Q^* > \underline{\Delta}$

(b) There exists $\overline{\Delta}(>0)$, such that the expected leftover inventory in the channel with a secondary market is smaller than without a secondary market if and only if $Q^* - Q^0 < \overline{\Delta}$

Let $\underline{Q} := Q^0 + \underline{\Delta}$ and $\overline{Q} := Q^0 + \overline{\Delta}$. Then when $Q^0 \le Q^*$, we will always have $\underline{Q} < Q^*$, and the expected sales in the channel with a secondary market will be greater than that without a secondary market, while the expected stockouts are lower. Also, the expected sales in the channel can be greater with a secondary market even when $Q^* < Q^0$ and the manufacturer sells less, as long as $Q^* > \underline{Q}$. When $\underline{Q} \le Q^* \le \overline{Q}$, the secondary market will lead to higher expected sales in the channel, with less expected stockouts and lower leftover inventory. Further, if $Q^0 \le Q^* \le \overline{Q}$, all three measures improve due to the secondary market.

3. Manufacturer's Intervention in Secondary Markets for Product Diversion

In Sect. 2 we have assumed that the secondary market is used solely by the retailers. We next investigate the impacts of the manufacturer's decision to intervene in the secondary market. In the absence of a secondary market, retailers sell their products to their primary-market customers over a single period, in a setting identical to the one described in Sect. 2. Given symmetric retailers, the optimal order quantity for each retailer i, Q^0, is therefore given in (1).

Suppose now that there exists a secondary market that serves a different customer base from the primary market. At the end of the first period the retailers can dispose of their excess inventory in this secondary market. Demand faced by each retailer in the secondary market is assumed to be characterized by a deterministic function of the form $a(1 - p_2/b)$, where p_2 is the secondary-market unit price and $a > 0$ and $b > 0$ are known constants.

It is further assumed that the secondary market demand level is independent of the primary market demand realization. Let Q be the quantity ordered by each retailer at the beginning of period 1, x_i (a draw from $F(\cdot)$) be the primary-market demand realized by retailer i, and $D_2 = na(1 - p_2/b)$ be the secondary-market aggregated demand given a unit price p_2. Using a methodology similar to the one described in Sect. 2, we obtain the equilibrium solution as follows:

Secondary-market unit price: Given any **Q** and **x**, total demand in the secondary market is given by $na(1 - p_2/b)$, whereas total supply is $\sum_{i=1}^{n}(Q - x_i)^+$. Let p_2^I be the limiting equilibrium price, given a retailers-only secondary market and an infinitely large n. p_2^I is determined by equating total supply with demand in the secondary market, which yields at the limit:

$$p_2^I = b\left[1 - \Gamma(Q)/a\right] \tag{6}$$

where $\int_0^Q F(x)dx := \Gamma(Q)$.

Primary-market order quantity: Each retailer chooses the quantity to order from the manufacturer so as to maximize her total expected profit from the primary and secondary markets, Π_R, which satisfies:

$$\Pi_R = \left\{\pi \int_0^Q x f_1(x)dx + \pi Q \int_Q^\infty f_1(x)dx - p_1 Q + p_2 \int_0^Q (Q - x)f_1(x)dx\right\} \tag{7}$$

The optimal order quantity is obtained by differentiating the total profit function (7) with respect to Q, which yields:

$$F(Q^i) = \frac{\pi - p_1 + \left(dp_2^I/dQ\right)\Gamma(Q^i)}{\pi - p_2^I}$$

Given the assumption of an infinitely large n, each retailer is a price-taker in the secondary market, i.e., $dp_2^I/dQ = 0$. Consequently, the above result can be simplified to:

$$F(Q^i) = \frac{\pi - p_1}{\pi - p_2^I}. \tag{8}$$

Q_i as specified in (8) will be optimal only as long as it results in non-negative p_2. Let $\overline{Q}$ be the value of Q that satisfies $\Gamma(Q) = a$, which, based on (6), yields $p_2' = 0$. If we denote by Q' the optimal order quantity, then Q' will be equal to Q_i only as long as $Q_i \leq \overline{Q}$.

Otherwise Q' will equal Q^0, the optimal order quantity with no secondary market.

By comparing (8) with (1), and given the assumption that the secondary market unit price p_2' cannot take on negative values, it is clear that the introduction of the secondary market will always lead the retailers to increase their initial order quantity Q.

THEOREM 4. *In equilibrium the secondary market will always increase the retailers' initial order quantity.*

With the introduction of the secondary market, the salvage value of any units available to the retailers at the end of the first period is increased from zero to $p_2' > 0$. A lower overage cost then provides an incentive for the retailers to increase their initial order quantity. Based on Theorem 4, it can be easily verified that the introduction of the secondary market will always increase the manufacturer's total profit as well as the primary market performance, measured by consumer sales and stockout level. That is, the introduction of the secondary market will benefit all entities along the supply chain: the manufacturer, retailers, primary-market customers, and secondary-market customers.

Next, suppose that the manufacturer decides to take part in the secondary market as a means for improving her profitability. To do so, the manufacturer may produce some additional quantity nK to be sold in the secondary market. It is assumed that the manufacturer decides on the value of K after the retailers place their orders, and that they are aware of the manufacturer's intention to enter the market ahead of time. Given unit production cost of c, for any Q, the manufacturer will determine $K(Q)$ so as to maximize total expected profits:

$$K(Q) = \arg\max_K \left\{ \Pi_M^{II} = n\left[-(Q+k)c + p_1 Q + p_2^{II} K\right] \right\} \qquad (9)$$

To solve this, we start with the secondary market to see how the choice of K affects the market price. Given any K, $\mathbf{Q}$ and $\mathbf{x}$, the equilibrium unit price in the secondary market, p_2^{II}, is determined by equating total demand with total supply, which satisfies at the limit case of a sufficiently large n:

$$p_2^{II} = b\left[1 - (K + \Gamma(Q))/a\right] \qquad (10)$$

Moving backward to (9), given $\mathbf{Q}$ and p_2^{II}, we find that the value of K that satisfies (9) is:

$$K^{II}(Q) = [a(b-c)/(2b) - \Gamma(Q)/2]^+ \qquad (11)$$

Let $\underline{Q}$ be the value of Q that satisfies $\Gamma(\underline{Q}) = a(b-c)/b$. Then based on (11), for all $Q > \underline{Q}$, $K^{II}(Q)$ should be set equal to zero. That is, when the retailers' initial order quantity Q is relatively high, so will be the expected level of excess inventory remaining for sale by the retailers in the secondary market, making it not beneficial for the manufacturer to participate in that market. Based on (10), it can be verified that for all $Q \leq \underline{Q}$, the resulting value of p_2^{II} will satisfy. $p_2^{II} > c$.

Finally, the initial order quantity to maximize the retailers' total expected profits satisfies:

$$F(Q^{ii}) = \frac{\pi - p_1 + (dp_2^{II}/dQ)\Gamma(Q)}{\pi - p_2^{II}}$$

Again, given the assumption of an infinitely large n and of price-taker retailers in the secondary market, the above result can be simplified to:

$$F(Q^{ii}) = \frac{\pi - p_1}{\pi - p_2^{II}} \qquad (12)$$

Q^{ii} as specified in (12) will be feasible and thus optimal only as long as it results in nonnegative values of p_2^{II} and $K(Q^{ii})$. Therefore, the optimal order quantity Q^{II} will be equal to Q^{ii} only if $Q^{ii} \leq \underline{Q}$ holds. Otherwise, it will be better for the manufacturer not to intervene in the secondary market, and consequently Q^{II} will be equal to Q^{I}.

Impacts of the manufacturer's intervention: Consider first the optimal order quantity Q. It turns out, as summarized in Theorem 5, that if the manufacturer decides to intervene in the secondary market, it will always lead the retailers to adjust their order quantity downwards:

THEOREM 5. *In equilibrium, $Q^{II} < Q^{I}$ will always hold.*

By offering additional units for sale in the secondary market, the manufacturer makes this market less attractive for the retailers, since it drives the expected unit price down. Consequently, it will be optimal for the retailers to reduce their initial order quantity Q.

Next we consider the impact of the manufacturer's intervention on her expected profit. The secondary market clearly provides an additional source of revenue for the manufacturer, however it also increases her production cost. Furthermore, the reduction in the retailers' order quantity also has a negative impact on the manufacturer's profits.

Comparing the expected profits of the manufacturer under the two scenarios, we obtain that the manufacturer will benefit from the decision to enter the secondary market if and only if:

$$\frac{1}{n}\left(\Pi_M^{II} - \Pi_M^{I}\right) = \left(Q^{II} - Q^{I}\right)\left(p_1 - c\right) + K^{II}\left(p_2^{II} - c\right) > 0 \qquad (13)$$

The result of (13) is intuitively clear: the manufacturer will benefit from intervening in the secondary market only as long as her total expected profits from the secondary market, $K^{II}(p_2^{II} - c)$, are sufficiently high to compensate for the reduction in profits from sales to the retailers at the beginning of the first period.

As for the retailers, it can be easily verified that the manufacturer's intervention will lower their expected profits. However, the manufacturer's intervention will never reduce the retailers' expected profits to a level lower than in the absence of the secondary market. At the same time, since the service level to the primary-market customers is only a function of the order quantity Q, it is clear from Theorem 5 that the manufacturer's intervention will always reduce the service level to this segment of the market, even though these customers will still be better off compared to the situation where the secondary market does not operate at all. Finally, the secondary market customers will always benefit from the manufacturer's intervention, as summarized in Theorem 6:

THEOREM 6. *In equilibrium, the manufacturer's intervention always reduces the equilibrium unit price and increases the market size of the secondary market.*

4.   The Informational Role of Secondary Markets

So far, we have considered the *operational* role of the secondary market when the demand distribution is accurately known at the beginning t_0 of a sales season. When the demand distribution is not perfectly known to retailers, the secondary market also plays an *informational* role (see Hayek, 1945). To investigate the informational aspects of the secondary market, consider the model in Sect. 2 with an informational twist where retailers start with a common normal prior distribution on demand mean, and update demand mean after observing the first period demands. We also allow the second-period selling price r_2 to be endogenously determined through the consumer market clearance condition, instead of being exogenously given.

The demand structure is modified accordingly. Retailer i's demand in each period has an additive noise and decreases linearly in the selling price. At the same selling price, the demands in the two periods are assumed to have the same mean for simplicity. That is, the demand for retailer i in period t ($t = 1, 2$) is $\widetilde{D}_{it} = \alpha - \beta r_t + \widetilde{\varepsilon}_{it}$, where α represents the upper bound of the mean demand, and $\widetilde{\varepsilon}_{it}$ follows the distribution $N(0, \sigma^2)$. $\widetilde{\varepsilon}_{it}$'s are i.i.d across retailers and the two selling periods. Denoting $\mu = \alpha - \beta r_1$ as the mean demand at selling price r_1, the demand faced by retailer i in period t can be rewritten as $\widetilde{D}_{it} = \beta(r_1 - r_2) + \mu + \widetilde{\varepsilon}_{it}$ All retailers, not knowing μ perfectly at t_0, have (the same) prior distribution $N(\upsilon, \tau^2)$. The prior predictive demand distribution is therefore given by ($N(\upsilon, \sigma_0^{\,2})$, where $\sigma_0^{\,2} = \sigma^2 + \tau^2$.

By t_1, the end of period 1, retailer i has observed his first-period sales x_i, and thus possesses a piece of private information. We consider two models that differ in the way that the private information is handled. We start with a Bayesian Updating (BU) model, described as a three-stage game:

1. At t_0, retailer i submits an initial order Q_i, determined by maximizing the expected profit in the two selling periods and in the secondary market. By symmetry, $Q_i = Q$ for each i, since we are interested in the symmetric equilibrium.
2. At t_1, retailer i updates his subjective demand distribution based on information set $I_i^0 = \{x_i\}$, using Bayesian statistics. Based on the updated demand distribution, he estimates the prevailing (consumer market clearance) selling price in the second period, and determines his optimal secondary inventory level by maximizing the expected profit in the secondary market and the secondary selling period. Each retailer observes the spot price in the secondary market, and buys/sells to reach the optimal inventory position for the second selling period. This step may be replicated multiple times, during which the secondary market price evolves endogenously to an equilibrium price, which clears the secondary market.
3. Retailers sell in the second period at the prevailing price that clears the consumer market.

Denote by G2 the subgame starting with the secondary market, and G the three-stage game given above. A Bayesian Updating (BU) equilibrium of the game G is given by the strategy profile $\{Q^b, q_i^b, p_2^b, r_{i2}^b\}$, where Q^b, q_i^b, p_2^b and r_{i2}^b are respectively, the optimal first-period order quantity, optimal secondary inventory level for retailer i, equilibrium secondary market price and the conjectured second period selling price by retailer i. Here, superscript b signifies the fact that the values are optimal for the BU model. Let $\mathbf{q^b} = \{q_1^b, q_2^b, \cdots, q_n^b\}$. Given Q^b, the BU equilibrium for subgame G2 is a tuple ($p_2^b, \mathbf{q^b}$) that fulfills the following conditions:

1. Retailer i's posterior on μ is formed with Bayesian statistics using set $I_i^0 = \{x_i\}$ That is, $\mu \mid x_i \sim N(m(x_i), k_1\tau^2)$, where
   $m(x_i) \equiv (\sigma^2 v + \tau^2 x_i)/(\sigma^2 + \tau^2)$, $k_1 \equiv \sigma^2/(\sigma^2 + \tau^2)$, and $k_2 \equiv 1 - k_1$.
2. Given p_2^b, retailer i finds it optimal to take inventory position q_i^b.
3. Given $\mathbf{q^b}$, the equilibrium price p_2^b clears the secondary market.

Our equilibrium for G is therefore defined as a symmetric (subgame-perfect Bayesian) Nash equilibrium; and the BU equilibrium for G2 is a Nash equilibrium in the secondary market. By solving the problem with backward-induction, we can prove the following theorem that characterizes the equilibrium for G:

THEOREM 7. *For n sufficiently large, a BU equilibrium of game G,* $\{Q^b, q_i^b, p_2^b, r_{i2}^b\}$, *fulfills the following system of equations:*

$$r_1 F_{v, \sigma_0}(Q^b) - G(Q^b) = r_1 - p_1, \tag{14}$$

$$q_i^b = m(x_i) + \beta(r_1 - r_{i2}^b) + \sigma_1 \Phi^{-1}(1 - p_2^b / r_{i2}^b), \tag{15}$$

$$\sigma_1 E_{x_{i|\mu}} \Phi^{-1}(p_2^b / r_{i2}^b(x)) = E_{x_{i|\mu}} \Gamma_{m(x), \sigma_1}(Q^b) - \Gamma_{\mu, \sigma_1}(Q^b), \tag{16}$$

$$r_{i2}^r(x_i) = r_1 + [m(x_i) - \Gamma_{m(x_i), \sigma_1}(Q^b)]/\beta] \tag{17}$$

where $\sigma_1^2 = \sigma^2 + k_1\tau^2$, and $G^b(Q) \equiv \int_0^\infty p_2^b(\mu) F_{\mu, \sigma} dF_{v, \tau}(\mu)$.

To calculate Q^b from (.14), it is important to know the functional relationship between p_2^b and μ, which is given by (16) implicitly. The following proposition explicitly states that p_2^b increases in μ under a very weak condition:

PROPOSITION 1. *Define* $\sigma_2^2 = \sigma_2^2 + k_2^2\sigma^2$ *Then* p_2^b *strictly increases in* μ *if the following condition is fulfilled:*

$$\Phi\left(\frac{Q-\mu}{\sigma}\right) - k_2\Phi\left(\frac{Q - k_1 v - k_2 \mu}{\sigma_2}\right) > 0.$$

Note that $Q > \mu$ is almost always true. As a result, condition (B) is extremely weak because $k_2 < 1$ and $\sigma_2 > \sigma$. We assume that it is fulfilled from now on.

The BU equilibrium given in Theorem 7 is not a stable one, since it is not "self-fulfilled." Note that from Proposition 1, $p_2^b(\mu)$ is a monotone function and is invertible. Suppose that the secondary market is currently at the BU equilibrium. A rational retailer, who observes p_2^b from the secondary market, is able to get an accurate estimate of μ because $p_2^b(\mu)$ is invertible. Such a retailer would have a desire to adjust his secondary inventory level based on the new estimate. When all retailers are rational, as retailers adjust their order quantities, the secondary market price continues to evolve, and the BU equilibrium cannot be sustained. If we assume that the secondary market price does eventually converge to an equilibrium, the relationship between p_2^b and μ cannot be described by the same (16) assumed by retailers earlier. The retailer's expected functional relationship between p_2^b and μ is therefore not fulfilled under this BU equilibrium.

We now develop a Rational Expectations (RE) model and apply it to the secondary market model. We show that the corresponding Rational Expectations equilibrium is self-fulfilled and therefore stable. Similar to the BU model, we use superscript r to denote the optimal quantities in the RE model. The key difference between the RE and the BU model is that retailer i makes his decision based on both x_i and p_2^r, the equilibrium secondary market price in the RE model. We give a formal definition of the RE equilibrium in the following:

DEFINITION 1. *Given* Q, *a rational expectations equilibrium of subgame $G2$ is a function* $p_2^r(\mu)$, *such that for any realization* $\bar{x} = \sum_{i=1}^{n} x_i / n$,

1. *The posterior distributions for demand mean are conditional on information set* $I_i = \{x_i, p_2^r\}$.
2. *The secondary inventory level, given the secondary market equilibrium price, maximizes individual retailer's expected profit based the posterior distribution.*
3. *The secondary market equilibrium price clears the secondary market.*
4. $p_2^r(\mu)$ *is monotone, and known by all retailers.*

We are interested in the case where n is sufficiently large so that $\bar{x}$ can be replaced by μ. The following theorem demonstrates the existence and uniqueness of the RE equilibrium in the secondary market:

THEOREM 8. *For n sufficiently large, the function* $p_2^r = [2\beta r_1 + \mu - \Gamma_{\mu,\sigma}(Q)]$ */ β is a RE equilibrium for subgame G2. In addition, there exists no other monotone function* $\hat{p}_2^r(\mu)$ *that is also a RE equilibrium.*

Since retailers already form their posterior distributions based on p_2^r, seeing the realization of p_2^r, retailers would have no desire to adjust their inventory levels from the RE equilibrium. The RE equilibrium is therefore stable, or self-fulfilled. The RE equilibrium for game G is given by another set of system of equations:

THEOREM 9. *For n sufficiently large, a RE equilibrium of game G,* $\{Q^b, q_i^b, p_2^b, r_{i2}^b\}$, *fulfills the following system of equations:*

$$r_1 F_{v,\sigma_0}(Q^r) - G(Q^r) = r_1 - p_1, \tag{18}$$

$$q^r = \mu + \beta(r_1 - r_2^r), \tag{19}$$

$$p_2^r = r_2^r / 2, \tag{20}$$

$$r_2^r = r_1 + [\mu - \Gamma_{\mu,\sigma}(Q^r)] / \beta \tag{21}$$

where $G^r(Q) \equiv \int_0^\infty p_2^r(\mu) F_{\mu,\sigma}(Q) dF_{v,\tau}(\mu)$.

Note that at the RE equilibrium, all retailers turn out to make the same (and correct) estimate on the second period selling price, and the optimal inventory levels are independent of each retailer's private information. The latter conclusion is a result of the assumption that $\tilde{\varepsilon}_{it}$'s are i.i.d, so observing p_2^r, a sufficient statistic of $\{x_1, x_2, \cdots x_n\}$, retailer i's private information is redundant. The optimality of the RE equilibrium can be demonstrated by considering a hypothetical Full Information (FI) model where after the first period, demand from all retailers are faithfully shared. It can be shown that the equilibrium given in Theorem 9 is also the unique equilibrium of the FI model. As a result, the RE equilibrium cannot be Pareto-improved. We therefore have demonstrated that the secondary market, among others, is a surrogate mechanism for sharing information among retailers.

5. Discussions and Concluding Remarks

Secondary markets are pervasive in various industries. For example, New York Stock Exchanges, NASDAQ and most of stock, options and futures markets in the world operate as secondary markets as investors buy and sell "initially publicly offered" securities. A vast literature exists in finance that studies these markets, but most of the research is set in pure exchange economy, with little in the production economy or in association with supply chain management. There exists a significant difference between the two economic settings. For example, trading of goods in the latter economy has direct impacts on the next steps of economic activities like sales and production. Shortage of goods entails permanent loss of production or sales opportunities, and surplus incurs inventory or spoilage costs, unlike the case of the exchange economy where only financial outcome is considered. Another difference is the existence of consumers in the latter economy. Demands come from consumers who will actually consume the goods, instead of reselling their inventory in the secondary market. Manufacturers, on the other hand, could supply more products to the secondary markets as we discussed in Sect. 3. As a result, a secondary market, or even primary markets in this regard, has different features and implications to the traders who can use it to avoid shortage or surplus of goods. It is natural to incorporate the "next step" in the analysis of secondary markets.

Two Roles of Secondary Markets

The three papers reviewed in the present chapter have a simple setting where retailers exchange goods among themselves at the ongoing price. A plausible scenario that fits the setting of Sects. 2 and 4 may be a capital equipment manufacturer who has several hundred clients. An example is Boeing and its large number of customers like commercial airlines and logistics service providers. To facilitate the convenience and operational efficiency of maintenance and repair of equipment, the manufacturer may open a secondary market where its clients can trade spare parts among themselves. According to Theorem 1, the manufacturer would prefer to exclude high-margin parts from the market trading. Also, if the secondary market is a separate market (e.g., an Internet auction market selling obsolete models for bargain searchers) like Sect. 3, then the manufacturer's inter-venetion would create channel conflicts. Retailers will buy less from the manufacturer, not because of sabotage, but because of the secondary market's less attractive margin (Theorem 5). But if the manufacturer does intervene, the winners will be the secondary market consumers who will face lower price (Theorem 6). Finally, the market may not only serve as a buffer to readjust his inventory position, but also as an informational source of market demand (Theorem 8 and remarks).

In summary, the secondary market is a driver of two forces of efficiency gain – *allocative* and *informational*. Allocative efficiency is achieved by the inventory pooling effect. While each retailer individually owns and manages a local inventory, the market serves as a middleman to redistribute the inventories across the local inventories. The decentralized inventory system with the market would attain the full efficiency of a centralized system, if the transshipment of goods were costless and instantaneous. Further, the market can aggregate diverse pieces of relevant information and present the summary in the form of price. In fact, the informational role of secondary markets puts the decentralized systems even better than the centralized system, since the latter has no easy way to obtain the information dispersed in bits and pieces across the channels (see Hayek, 1945).

While most markets play both allocative and informational roles, we still find some markets dedicated to one role only. For example, Toyota asks its Japanese dealers to forecast their local demand and place orders 2 months ahead of delivery time (Lee et al., 2004). At this time dealers give only partial specs of the cars being ordered. Later, 10 days before the production of the order, dealers give the remaining specs. Even later, 3 days before the actual production dealers can request (although not always accepted) spec changes in a limited way. This postponement method allows dealers to update their local market forecast as they get closer to the delivery date. But such flexibility would not be sufficient to the dealers since the market demand continuously and abruptly changes. One additional feature of the Toyota distribution system is the Locator Program (in Japan) or the Dealer Daily system (in the US) that serves as a secondary market where dealers can swap among themselves the cars ordered or delivered. Interestingly, trade is done at the actual *dealer cost*, and dealers usually swap cars one for one after adjusting the cost differential. Here the market plays only the allocative role, without any informational role.

The opposite case also exists where the market is purely informational. According to Chen and Plott (2002), Hewlett–Packard opened an experimental Information Aggregation Mechanism (IAM), a market where "sales-forecast" state-contingent securities are traded. The market operates as follows. About 3 months before the selling season, 20–30 managers (who are experts in the market forecast domain) are invited to participate in the "stock" market. One share of stock called "A101–200Sept" will give its holder \$1 on a closing date if the state "sales of product A in September falls between 101 and 200" is actually realized. There are other competing stocks (like A0–100, A201–300, etc), so that covered states are mutually exclusive and complete. On the first opening day of the market each participant starts with a certain endowment of shares of different stocks, and over time trades them in the IAM market at ongoing market prices. As traders receive information from diverse private and public sources and update their demand forecasts, they buy or sell different shares, and the market prices adjust to the popularity (i.e., demand and supply) of different stocks. At the end of the experiment, they found that the market price (volume-averaged over the last days of trading) beat the internal forecasts in six out of eight cases. Note in this case that the market is purely for the purpose of collecting information, and no allocation of goods is directly involved. Supply chains often suffer from informational distortion through shortage gaming (Lee et al., 1997a, 1997b), but the market offers a chance to rectify it and even enhances it to a consolidated form.

Conditions of Successful Secondary Markets for SCM

Building a successful B2B exchange is not easy. While B2B exchanges (primary or secondary) offer a clear value proposition from a theoretical perspective, not all exchanges have fared well. In addition to a well-designed trading mechanism, the prerequisites of a successful B2B exchange also include: (a) standardized and well-specifiable products (Rangan, 1999), (b) quality assurance mechanisms (e.g., the quality certification program offered by http://www.ironplanet.com for used building equipment), (c) fraud prevention, protection, insurance or dispute resolution (e.g., eBay), (d) privacy protection (e.g., fastparts), (e) easy-to-use trading system (e.g., Marketscape at Caltech), (f) low transaction costs, and (g) a large population of motivated and informed traders. Regarding the last point, a large number of traders do not only create an efficient market, but also provide privacy and protection against collusion and market cornering.

Limits and Extension of the Model

The model we discussed in the chapter is very limited in its scope and setup. First, only two forms of secondary markets are considered: Sections 2 and 4 assume a secondary market where traders directly exchange "goods" among themselves, and Sect. 3 assumes a secondary market that targets a different and independent market, to whom retailers, and possibly the manufacturer too, sell their excess goods. But there are alternative markets or modes of operation for SCM. For example, the rights to the goods may be "securitized" in the form of futures and forwards. In fact, DRAMeXchange.com currently offers trading on DRAM futures. The difference between direct exchanges and "securitized" exchanges lies in that goods trading is (at least implicitly) coupled with the delivery requirement from the seller to the buyer. In the futures market, however, one can buy and resell the rights to others, who may do the same, until the specified delivery date. Accordingly, the latter (free of transportation burdens) is more conducive to active trading not only for traders' own use, but also for stockpiling, diversion and speculation. Note also that several third-party foundry companies like TSMC or UMC securitized "capacity," too. By owning the capacity stock, a fabless chip-maker can secure a certain level of manufacturing capacity in anticipation of shortage during the boom time. More research may be in need to study the advantages and drawbacks of this "security" approach, compared with goods trading or other contractual forms, as well as its impact on the supply chain.

Second, the model is an equilibrium analysis and suppresses all the details of the equilibrium formation process. Although we have demonstrated the existence and uniqueness of market equilibriums in different settings, there could be multiple paths that lead to such equilibriums. When the demand distribution is known, retailers can theoretically predict the secondary market equilibrium price using (4), and one can expect that the secondary market price quickly converges to the equilibrium value. A simple path that leads to the secondary market equilibrium is for retailer i to buy/sell $\left| q * - \left(Q - x_i \right)^+ \right|$ units from/to the market at the equilibrium price. When the demand distribution is not known perfectly, one interesting path that might lead to the RE equilibrium is to assume that retailers trade multiple rounds, the trading price in each round is public information, and retailers are only "boundedly rational," i.e., they use Bayesian updating after each round based on trading prices in previous rounds (see Cyert and Degroot, 1974). In reality, the boundary between different rounds is blurred, but the fact that each retailer could trade many times is important to the RE equilibrium formation.

On a related note, the model assumes symmetric traders – all starting with the same level of confidence in the demand forecast. But in reality there may be significant differences in their forecasting accuracy. If so, how would the market differentiate the forecasters and reach a correct equilibrium? To elaborate, consider the following new setting with asymmetric traders. There are n forecasters facing the same unknown parameter θ (mean demand). Forecaster i observes a private signal $\overline{\theta}_i$ that is a noisy unbiased estimate of θ. The accuracy of the estimate differs across forecasters, so that we assume $\overline{\theta}_i \sim N(\theta, \sigma_i^2)$, where the distribution of σ_i^2 is known to all. Define the precision γ_i^2 as the inverse of the variance; i.e., $\gamma_i^2 = 1/\sigma_i^2$. If some omniscient central planner could observe all the signals $\overline{\theta} = \left(\overline{\theta}_1, \overline{\theta}_2, \cdots, \overline{\theta}_n \right)$, she may consider the linear estimate θ^* that minimizes its square error by solving:

$$\min_{\theta} E\left(\theta - \alpha\overline{\theta} \right)^2$$
$$\text{subject to } \alpha \bullet 1 = 1$$

Straightforward algebra shows that the minimum variance unbiased linear estimate is $\theta^* = \alpha^* \overline{\theta}$, where $\alpha_i^* = \gamma_i^2 / \sum \gamma_j^2$. Thus, the weight to a signal is proportionate to its precision, and this makes intuitive sense. In particular, if one signal has a precision value close to infinity (or close-to-zero variance), the weight on that signal would converge to 1, as it should. In fact, one can further show that this linear statistic is the unique uniformly minimum variance unbiased (U.M.V.U.) estimate – the most viable qualification as a parametric estimate (see Bickel and Doksum (1977)).

Now we consider two models of market to investigate how the market differentiates different forecasters and aggregates hidden information without involving an omniscient central planner. In the first model, consider an *exchange* economy where individual forecasters act as traders in a stock market. A fixed number of shares is available and traded in the market at an ongoing price p. Without loss of generality, total supply is one, and it is infinitely divisible. One share will return X at the end of the trading day. Trader i is risk averse facing an exponential utility function. His decision is the number β_i of shares to maximize his expected utility. Thus, the objective of trader i is given by

$$\max_{\beta_i} E(\beta_i X) - \frac{A}{2} Var(\beta_i X) - \beta_i p$$

where A is a measure of risk aversion that is the same across all traders. The return X is drawn from $N \sim (\mu, \sigma^2)$. No trader knows θ or σ^2, but trader i privately observes $\overline{\theta}_i$ and σ_i^2. Note in particular that the precision γ_i^2 of his observed estimate is different across traders but privately known to the trader himself. However, the distribution of σ_i^2 (and γ_i^2) is common knowledge. Then, to trader i, X is viewed as drawn from $N \sim (\overline{\theta}_i, \sigma_i^2)$.

Thus, he solves:

$$\max_{\beta_i} \beta_i \overline{\theta}_i - \frac{A}{2} \beta_i^2 \sigma_i^2 - \beta_i p,$$

which yields $\beta_i^* = \dfrac{\overline{\theta}_i - p}{A \sigma_i^2}$. In equilibrium it should hold:

$\sum \beta_i^* = \sum \dfrac{\overline{\theta}_i - p}{A \sigma_i^2} = 1$. Finally, the equilibrium price should satisfy

$p^* = \dfrac{\sum \overline{\theta}_j \gamma_j^2 - A}{\sum \gamma_j^2}$. Since the denominator is common knowledge, the market price reveals the value $\dfrac{\sum \overline{\theta}_j \gamma_j^{\,2}}{\sum \gamma_j^{\,2}}$, which is identical to the U.M.V.U.

estimate (i.e., $\theta^* = \alpha \overline{\theta}$). Note that the market aggregates scattered information and places a proper weight to each estimate based on the precision. This is in contrast to Obermeyer's forecast panel approach (Hammond and Raman, 1994), where the forecast is the equally weighted average of participants' votes, regardless of each participant's confidence level.

Now turn to the second model that is a simplified version of the secondary market as discussed in Sects. 2 and 4. In the second period there is only one unit left to share among n retailers. Retailer i's forecast is again given by $N \sim \left(\overline{\theta}_i, \sigma_i^{\,2}\right)$, as in the previous exchange model. Its optimal

stock level would be $S_i = \overline{\theta}_i + k\sigma_i = \overline{\theta}_i + \Phi^{-1}\left(\dfrac{\pi - p}{\pi}\right)\sigma_i$, where p is the

equilibrium market price. It is straightforward to show that the price is given

by $p = \pi\left[1 - \Phi\left(\dfrac{1 - \sum \overline{\theta}_i}{\sum \sigma_i}\right)\right]$. Since the distribution of variance is common

knowledge, the market price is a function of the equally-weighted average ($\sum \overline{\theta}_i / n$) of forecasts. This estimate is unbiased and consistent, but not of

minimum variance. In particular, if one retailer i has infinite precision (i.e., zero variance), the hypothetical omniscient central planner would recognize θ_i as the true mean demand, but the market price this time would miss its significance by treating it equally with other forecasts. Hence, we face the possibility that the market may not be so efficient in aggregating scattered information. As a possible reason, note that the newsvendor model induces an unsure retailer to carry higher inventory (when underage cost is higher than overage cost) – contrary to what the market would desire for efficient information gathering.

One compromise to the dilemma would be the possibility that the high-precision retailer would find an arbitrage opportunity from the discrepancy between the current market price (in disequilibrium) and his accurate assessment, and trade to make profits. It will expedite the equilibrating process, and efficiency may be regained. This is admittedly a temporary fix for an important issue, and further investigation is called for. We leave it for future research.

References

Bickel, P.J. and Doksum, K.A., 1977, Mathematical Statistics, olden-Day, Inc.

Chen, K.Y. and Plott, C., 2002, Information Aggregation Mechanisms: Concept, Design and Implementation for a Sales Forecasting Problem, *Social Science Working Paper* #1131, Division of Humanities and Social Sciences, California Institute of Technology, Pasadena, California 91125, March 2002.

Cyert, R. and DeGroot, M., 1974, Rational Expectations and Bayesian Analysis, *J. of Political Economy*, 82 (3): 521–536.

Eppen, G. and Schrage, L., 1981, Centralized Ordering Policies in a Multi-warehouse System with Lead Times and Random Demand, in L.B. Schwarz Ed., *Multi-Level Production/Inventory Control Systems: Theory and Practice, TIMS Studies in the Management Sciences*, 16, North-Holland, Amsterdam, pp. 51–68.

Gallien, J. and Wein, L., 2000, Design and Analysis of a Smart Market for Industrial Procurement, *Working Paper*, Sloan School of Management, MIT.

Gerchak, Y., Magazine, M.J., and Gamble, A.B., 1988, Component Commonality with Service Level Requirements, *Management Science*, 34 (6): 753–760.

Hammond, J. and Raman, A., 1994, Sport Obermeyer, Ltd, *Harvard Business School case*, pp: 9-695-022.

Hayek, F.A., 1945, The Use of Knowledge in Society, *American Economic Review*, 35: 519–530.

Lee, H.L., 1987, A Multi-echelon Inventory Model for Repairable Items with Emergency Lateral Transshipments, *Management Science*, 33 (10): 1302–1316.

Lee, H.L., Peleg, B., and Whang, S., 2004, Toyota: Demand Chain Management, Stanford case, Graduate School of Business, Stanford University.

Lee, H.L., Padmanabhan, V., and Whang, S., 1997a, Information Distortion in a Supply Chain: The Bullwhip Effect, *Management Science,* 43 (4): 546–558.

Lee, H.L., Padmanabhan, V., and Whang, S., 1997b, The Bullwhip Effect in Supply Chains, *Sloan Management Review*, 38 (3) Spring, 93–102.

Lee, H.L., and Whang, S., 2002, The Impact of the Secondary Market on the Supply Chain, *Management Science*, 48 (6): 719–731.

Milgrom, P., and Weber, R.J., 1982, A Theory of Auctions and Competitive Bidding, *Econometrica*, 50 (5): 1089–1122.

Peleg, B., and Lee, H.L., 2002, Secondary Markets for Product Diversion with Potential Manufacturer's Intervention, *Working Paper*, Stanford University.

Rangan, V.K., 1999, *FreeMarkets Online*, Harvard Business School case, 9-598-109.

Whang, S., and Zou, Y., 2003, The Informational Role of the Secondary Market in a Supply Chain, *Working Paper*, Stanford University.

Global Diffusion of ISO 9000 Certification Through Supply Chains

Charles J. Corbett
UCLA Anderson School of Management, USA

Abstract: The ISO 9000 series of quality management systems standards is widely diffused, with over 560,000 sites certified in 152 countries (as of December 2003). Anecdotal evidence suggests that global supply chains contributed to this diffusion, in the following sense. Firms in Europe were the first to seek ISO 9000 certification in large numbers. They then required their suppliers to do likewise, including those abroad. Once the standard had thus entered other countries, it spread beyond those firms immediately exporting to Europe to be adopted by many other firms in those same countries. This paper empirically examines the validity of this view of the role of supply chains in global diffusion of ISO 9000. To do so, we decompose the statement that "supply chains contributed to the global diffusion of ISO 9000" into a series of four requirements that must be met in order for the original statement to be supported. We then use firm-level data from a global survey of over 5,000 firms in nine countries to test the hypotheses that correspond to these requirements. Our findings are consistent with the view that ISO 9000 did diffuse upstream through global supply chains. In short, this means that firms that export goods or services to a particular country may simultaneously be importing that country's management practices. We conclude by suggesting how these findings might form the basis for future research on the environmental management systems standard ISO 14000.

Key words: ISO 9000; ISO 14000; quality management; supply chains; diffusion; global; empirical; survey.

1. Introduction

Do management practices diffuse through global supply chains? That is the question underlying this research. There are many mechanisms by which management practices might spread: through industry associations, government interventions, consulting firms, management education, etc. Some of these mechanisms operate primarily within countries, others could contribute to practices spreading across countries. Our interest here is in the question whether management practices diffuse across countries as a result of global supply chains. In other words, when firms export goods or services to a country, do they also import some management practices from that country?

We study this question in the context of ISO 9000, the widely diffused quality management systems standard. ISO 9000 was introduced in 1986 and had been adopted by over 560,000 sites in 152 countries by December 2003 (ISO, 2004). Anecdotal evidence suggests that this worldwide spread occurred in part through supply chains. According to that view, firms in Europe were early to adopt ISO 9000 in large numbers; they then imposed the standard on their suppliers, including those abroad. These foreign suppliers essentially imported the standard into their own country, after which it spread to other firms in the same country.

This paper empirically examines this view of global diffusion of ISO 9000. To do so, we decompose the statement that "supply chains contributed to global diffusion of ISO 9000" into a series of four requirements that must be met in order for the original statement to be supported. This gives a self-contained theoretical framework for studying diffusion in supply chains; we relate this to the well-known Bass (1969) model of diffusion, which combines a constant innovation rate with a contagion effect, in Appendix A. Applying this theoretical framework to the case of ISO 9000, the four requirements translate into specific hypotheses. We combine global country-level certification data with firm-level data collected in a survey of over 5,000 firms in nine countries to test these hypotheses. Overall, we find strong support for the view that ISO 9000 was initially adopted in large numbers by firms in Europe, then "imported" by supplier firms in other countries exporting to Europe. Those supplier firms, in turn, trigger traditional single-market diffusion mechanisms, hence contributing to certification by yet other firms within their country. This is consistent with Dekimpe et al.'s (2000b) view that, in studying diffusion, both breadth (across countries) and depth (within countries) must be considered simultaneously. We conclude that supply chain pressures related to export flows did contribute to global diffusion of ISO 9000.

Below, in Sect. 2, we review selected literature on ISO 9000 and related standards and on diffusion theory. Section 3 presents our theoretical framework. Section 4 describes the data, and Sect. 5 discusses the methodology and results. Section 6 summarizes our findings and the limitations of this study.

A good concise introduction to ISO 9000 is available at http://www.iso.org, and states, among others: "ISO 9001:2000 specifies requirements for a quality management system for any organization that needs to demonstrate its ability to consistently provide product that meets customer and applicable regulatory requirements and aims to enhance customer satisfaction,"[29] where ISO 9001:2000 is the revised version of the standard, introduced in 2000, integrating the earlier ISO 9001, 9002 and 9003 standards, which were first introduced in 1986. Certification to ISO 9000 means that a third-party auditor has verified that the firm's quality management system complies with the requirements of the standard, and this certification must be renewed every 3 years. We will refer to the year in which the certification is first awarded as the year of adoption, although, strictly speaking, the preparations for a first certification can take 6–18 months. We will also occasionally refer to ISO 14000, the environmental management systems standard, introduced in 1996.

2.   Literature

This paper draws on and contributes to several literatures: that on management practices and standards in general and on ISO 9000 and ISO 14000 certification in particular, and that on global diffusion processes. Some of the recent theoretical perspectives on ISO 14000 are relevant for ISO 9000 too, which is why we include them here.

Early academic surveys on ISO 9000 include Brown et al.'s (1998) study of small firms' experiences with ISO 9000 and Lee's (1998) survey of ISO 9000 in Hong Kong. Terziovski et al. (1996) find no link between ISO 9000 and organizational performance among Australian firms. Anderson et al. (1999) use COMPUSTAT data to find that exports to Europe and elsewhere increase US firms' likelihood of seeking ISO 9000, but they do not distinguish between early and late adopters as we do here. Terlaak and King (2005) find sales growth that is consistent with signalling models contributing to adoption of ISO 9000. Naveh and Marcus (2004), using a detailed survey of ISO 9000 in the US, find that "going beyond" the

---

[29] See http://www.iso.org/iso/en/iso9000-14000/understand/selection_use/selection_use.html; last accessed August 18, 2005.

requirements of the standard increases its value to the firm, and Naveh et al. (2004) find no difference between early and later adopters with respect to external pressure to seek certification. Delmas (2002) finds that institutional theory (Scott, 1995) contributes to explaining early adoption of ISO 14000.

Some recent work has focused explicitly on global diffusion of ISO 9000 and ISO 14000. Mendel (2001) discusses various mechanisms driving global spread of ISO 9000. Guler et al. (2002) use national ISO 9000 certification levels to find that several forces, including trade relationships, drive global diffusion, while Delmas (2003) uses national ISO 14000 certification levels to find that reasons related to cost minimization and to legitimacy both contribute to adoption. Corbett and Kirsch (2001) show that ISO 14000 certification levels by country depend on ISO 9000 certification levels, export-propensity, and environmental attitudes. Christmann and Taylor (2001) find that foreign ownership and exports to more developed countries increase the likelihood of ISO 14000 certification among Chinese firms. Neumayer and Perkins (2004) use a panel data study of 142 countries to find that certifications per capita are positively correlated with, among others, stock of foreign direct investment and exports to Europe and Japan. Potoski and Prakash (2004) find that factors driving adoption of ISO 14000 within a country include the degree of adoption in export partners and the degree of regulatory flexibility in the country. Albuquerque et al. (2005) modify the Bass (1969) diffusion model to include cross-country effects and, using Bayesian estimation techniques, find that bilateral trade and geographical proximity are the strongest drivers of cross-country diffusion of ISO 9000, while cultural similarity also plays a role for ISO 14000.

The literature on diffusion of management practices is large; see, for instance, Dekimpe et al. (2000b) for a review of the limited research to date on global diffusion. Lücke (1996) finds that the level of economic development of countries affects the timing of the start of the diffusion process for weaving technology, but not the diffusion rate itself. Various other factors that drive diffusion of management practices are proposed by Abrahamson (1991) and subsequent work.

None of this work has directly asked adopters about the role of supply chains in explaining global diffusion. The contribution of the current paper is to formally state the requirements for claiming that supply chains contribute to global diffusion, and to apply this framework to ISO 9000 and test it. In doing so, we make use of the extensive literature on survey research methods, especially that dealing with international mail surveys, reviewed in Singh (1995) and Harzing (1997).

3.   Theoretical Framework

Consider a global economy, with firms in each geographic region trading goods and services with other firms, in the same region and elsewhere. A new management practice (in our case, ISO 9000) is introduced somewhere in this global economy, and gradually adopted by firms in many regions. A naive way of linking this global diffusion to supply chains would be to simply correlate the diffusion patterns observed with the deepening trade links between countries. Clearly, such an approach would not establish causality and would fail to account for several possible alternative explanations. To rule out such alternative explanations, we decompose the statement that "supply chains contributed to global diffusion" into a series of four requirements. Below we outline these requirements; later we use them to formulate and test specific hypotheses. In Appendix A, we describe how an adaptation of the Bass model of diffusion (Bass, 1969) can be used to generate largely identical hypotheses. At no time will we argue that global diffusion occurred exclusively as a result of supply chain interactions; however, if the four requirements below are met, there is strong evidence that supply chain interactions contributed to global diffusion.

The first requirement (R1) to support the view that global supply chains contributed to global diffusion of ISO 9000 is that there must exist considerable heterogeneity in timing of adoption across geographic regions. Without such chronological heterogeneity, it would be meaningless to claim that ISO 9000 spread from one region to another at all, let alone that any particular mechanism contributed to that spread. If this first requirement is met, we can distinguish between early- and later-adopting regions.

Assuming that this first requirement is met, the second requirement (R2) is that firms in the early-adopting regions must exert more pressure on their immediate suppliers to seek ISO 9000 certification than firms in later-adopting regions. Without this second requirement, there would be no supply chain-related causal link between diffusion patterns in early- and later-adopting regions. This pressure can take several forms. It can be explicit, e.g., by excluding non-certified firms from bidding for supply contracts (as is increasingly the case with ISO 9000). It can be implicit, e.g., by including questions about certification status in vendor selection questionnaires (as is currently becoming more common for ISO 14000). It can also be indirect, e.g., through the perception held by many non-European firms that ISO 9000 certification is somehow a requirement for exporting to the European Union. Any of these mechanisms would put pressure on firms in later-adopting regions to seek certification.

If the first two requirements are met, exporting firms in later-adopting regions experience pressure from customers in early-adopting regions to seek certification. The third requirement (R3) is that these exporters are indeed the first to seek certification within their (later-adopting) region. At that point, we can say that firms that export goods and services into another region imported ISO 9000 from that region. This third requirement addresses the breadth of diffusion from early-adopting to later-adopting regions (Dekimpe et al., 2000b). If the first two requirements are met but not the third, adoption of ISO 9000 in other regions would not be a result of the pressure exerted by customer firms in the early-adopting regions. Note though that the causality can take either direction: exporting firms may seek certification in order to protect their exports, or firms that seek certification may find themselves able to export more. Both are consistent with the supply chain diffusion perspective so we do not need to specify a direction of causality for the third requirement.

The fourth and final requirement (R4) addresses depth of diffusion within the later-adopting regions: firms that adopt later must be less motivated by export-related factors than early-adopting firms. In other words, ISO 9000 spreads to other firms beyond the early-adopting exporting firms through traditional single-market diffusion mechanisms. This could be through pressure from exporting firms on their domestic suppliers, through professional societies, trade organizations, word-of-mouth, etc. Without this fourth requirement, the spread of ISO 9000 in later-adopting areas would be limited to exporting firms, hence failing the depth requirement.

If these four requirements are satisfied, one can conclude that global supply chains contributed to global diffusion of ISO 9000, as depicted in Fig. 1. Other factors undoubtedly do also contribute to global diffusion of ISO 9000; for instance, professional societies and word-of-mouth also cross borders. However, if such traditional, single-market diffusion mechanisms fully explained global diffusion of ISO 9000, several of the four requirements above would not be met. For instance, if adoption of ISO 9000 in one of the later-adopting regions was the result of a government policy encouraging certification, R2 and R4 would not be met. Early- and later-adopting firms should, in that case, not differ in the degree to which they were motivated by export-related factors, hence failing R4. If the government policy first targeted exporting firms, R3 might be satisfied, but in that case one could argue that the government certification policy was merely a response to already existing supply chain pressures; otherwise, why focus on exporting firms? If that supply chain pressure did not already exist, R2 would not be met. A similar reasoning can rule out level of economic development as sole driving force of diffusion once all four requirements are met.

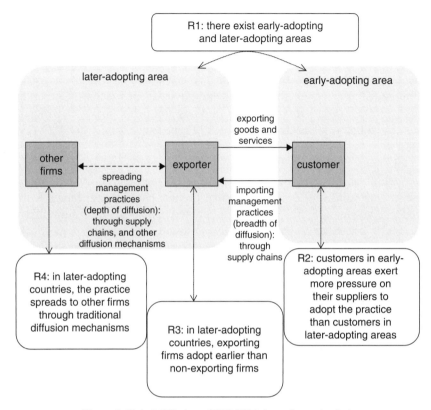

Figure 1. Global diffusion of ISO 9000 through supply chains

If diffusion follows a strictly geographical path, we would find a non-random pattern of early- and late-adopting regions in R1, but no effects for R2, R3 and R4. If cultural ties between countries were the main driver of diffusion, R3 may be satisfied if exporting firms are more exposed to other cultures, but R2 and R4 would not be met. Finding support for all four requirements enables us to rule out alternative explanations with some confidence and conclude that supply chains did contribute to global diffusion of ISO 9000. Hence, these requirements form a concise and falsifiable basis for testing whether global supply chains contributed to global diffusion of ISO 9000.

Next, we present the data used in this study, after which we reformulate the four requirements as precise hypotheses to be tested.

4.   Data

Data from several sources were used in this study. During 1997–1999 we conducted open-ended interviews in Brazil, Japan, the Netherlands, Taiwan, Uruguay, and the US, with firms with ISO 9000 and/or ISO 14000 certification, registrars, accreditation bodies and government agencies. Corbett and Kirsch (2000, 2001) summarize the findings from this phase. Certification levels by country for each year since 1993 were obtained from ISO (ISO, 2003).

Our main source of data is a mail survey, conducted in 17 countries. After eliminating countries with insufficient responses, nine countries remain: Australia/New Zealand, Canada, France, Hong Kong, Japan, Korea, Sweden, Taiwan, and the US. In each country, the survey was administered by local researchers; Harzing (1997) found that such domestically-administered surveys achieve higher response rates. A pilot survey among 200 US firms during the summer of 1999 yielded 23 responses. After slight modification, the same basic survey form was used in all countries, translated by the local partner. In total, 5,295 observations were used in this analysis. Response rates, shown in Table 1, varied from 8.8 to 50.4%. These figures compare favorably with Harzing's (1997) finding that response rates for international mail surveys among an industrial population are typically in the 6–16% range. The survey questions used here are shown in Appendix B. In questions 6 and 7, it would be impossible to ask firms about the importance of and the pressure exerted by customers in each of 100 or more individual countries, hence the aggregation into larger regions. The motivations included in question 9b are based on findings from earlier surveys on ISO 9000 and/or ISO 14000 certification (see the literature review) and on our interviews.

Table 1. Response rates by country

country	questionnaires mailed	usable responses received	response rate
Australia / New Zealand	3000	611	20.4%
Canada	561	198	35.3%
France	2000	445	22.3%
Hong Kong	1200	131	10.9%
Japan	5000	2261	45.2%
Korea	1361	120	8.8%
Sweden	268	135	50.4%
Taiwan	2142	455	21.2%
US	5000	939	18.8%
Total	**20,532**	**5,295**	**25.8%**

Care was taken to avoid the pitfalls of international mail surveys, discussed in Singh (1995). These include country biases due to cultural differences in response behavior or due to differences in sampling procedures. Some differences in sampling procedures were inevitable, due to lack of national databases of ISO 9000 and/or ISO 14000 certifications. In the US, certification data were obtained from WorldPreferred in Toronto. Five thousand forms were mailed in Winter 2001, to all (approximately 1,100) ISO 14000 certified firms and the remainder to randomly selected ISO 9000 certified firms. Monetary and non-monetary incentives were used: a postcard from the author's university was included in the survey, to which a "golden dollar" US$1 coin was attached, which presumably contributed to the relatively high 18.8% response rate. A telephone follow-up among 100 non-respondents in November 2001 yielded 37 responses; these exhibited similar breakdown by industry and number of employees worldwide as the original respondents, but had lower global sales. No material non-response bias was found in the questions included in the follow-up, such as year of certification and overall benefits of certification.

In Japan, national certification data were obtained from the Japan Accreditation Board (JAB). A cover letter from JAB was included with the survey, but no incentive. In France and Korea, addresses of certified companies were obtained from leading registrars. In addition to being inevitable, Keown (1985) found that such flexibility in sampling practices is in fact desirable.

We did notice some country biases; for instance, Taiwanese respondents reported higher scores on all motivation and benefit categories than other respondents. To avoid country bias, our analysis does not compare means across countries. Only for requirement R2 do we aggregate data across countries (Tables 8 and 9), but there too we report separate results by country (Table 10). Singh's (1995) concern with separate analyses by country is that of maintaining the overall Type-I error rate. This is not a problem in our case, as most of the results are significant at the 1% or even stricter significance levels.

Table 2 provides descriptive statistics on the demographics of the respondents, while Tables 3–5 show correlations between responses to the survey questions used here.

Table 2. Descriptive statistics: demographics

	N	mean	median	st.dev.	min.	max.
Q1b: employees worldwide (7-point scale)	4911	3.53	3	1	7	1.69
Q5a: year of initial ISO 9000 certification	5118	1997.21	1998	1983	2001	2.53

Q3: ownership	% of respondents
publicly owned	16.06%
privately owned	75.17%
state-owned	2.54%
foreign ownership	8.97%

Q4a: nature of business	% of respondents
services	25.91%
manufacturing	59.19%
construction	14.96%
software	1.65%

Q4b: activity (manufacturing)	% of respondents
communications	3.12%
computer eq. or peripherals	3.58%
electronics	9.73%
semiconductors	1.61%
mechanical	15.48%
automation	2.22%
food	4.32%
plastic	7.36%
chemical	8.25%
textile	1.74%
metal	14.84%
pharmaceutical	3.12%
others	3.58%

Table 3. Correlations for survey question 6: location of immediate customers

	emp.	year	Europe E	Japan J	North America NA	South America SA	Australia / New Zealand AZ	Asia w/o Japan AS	Africa A
mean	3.53	1997.21	2.38	3.15	2.54	1.66	1.73	2.47	1.20
st.dev.	1.69	2.53	1.61	1.78	1.75	1.09	1.38	1.56	0.62
N	4911	5118	5026	5026	5026	5026	5026	5026	5026
emp.	1.00	-0.35	0.28	-0.01	0.35	0.29	0.01	0.27	0.15
year		1.00	-0.26	0.36	-0.33	-0.24	-0.40	-0.31	-0.23
E			1.00	-0.05	0.52	0.57	0.13	0.31	0.34
J				1.00	-0.05	0.13	-0.26	0.12	-0.01
NA					1.00	0.63	0.20	0.31	0.39
SA						1.00	0.26	0.44	0.51
AZ							1.00	0.12	0.52
AS								1.00	0.28
A									1.00

Table 4. Correlations for survey question 7a: proportion of customers requiring ISO 9000 certification

	emp.	year	Europe (E)	Japan (J)	North America (NA)	South America (SA)	Australia / New Zealand (AZ)	Asia w/o Japan (AS)	Africa (A)
mean	3.53	1997.21	3.37	3.08	3.12	2.43	2.77	2.76	1.94
st.dev.	1.69	2.53	1.24	1.11	1.24	1.21	1.32	1.17	1.15
N	4911	5118	2310	3126	2344	1499	1131	2518	532
emp.	1.00	-0.35	0.10	0.05	0.09	0.09	0.04	0.07	0.13
year		1.00	-0.15	-0.04	-0.12	-0.08	-0.15	-0.17	-0.19
E			1.00	0.64	0.71	0.62	0.66	0.61	0.49
J				1.00	0.62	0.60	0.71	0.68	0.59
NA					1.00	0.69	0.61	0.59	0.50
SA						1.00	0.69	0.68	0.75
AZ							1.00	0.64	0.68
AS								1.00	0.64
A									1.00

Table 5. Correlations for survey question 9b: motivations for seeking ISO 9000 certification

	emp.	year	cost reductions (Q9b1)	quality improvements (Q9b2)	marketing advantage (Q9b3)	customer pressure / customer demands (Q9b4)	many competitors were already ISO 9000 (Q9b5)	benefits experienced by other certified (Q9b6)	avoid potential export barrier (Q9b7)	capturing workers' knowledge (Q9b8)	relations with authorities (Q9b9)	relations with communities (Q9b10)	corporate image (Q9b11)
mean	3.53	1997.21	3.08	4.23	3.81	3.42	2.84	2.39	2.40	3.37	2.60	2.54	4.02
st.dev.	1.69	2.53	1.06	0.81	0.97	1.16	1.17	0.99	1.29	1.06	1.19	1.13	0.87
N	4911	5118	4890	5040	5000	4995	4968	4912	4873	4963	4909	4930	5015
emp.	1.00	-0.35	0.00	-0.02	0.00	0.08	0.11	0.08	0.19	-0.03	-0.04	-0.05	0.00
year		1.00	0.02	0.08	-0.14	-0.17	-0.06	-0.14	-0.29	0.13	0.00	0.07	0.00
Q9b1			1.00	0.42	0.11	0.04	0.07	0.29	0.16	0.36	0.20	0.29	0.20
Q9b2				1.00	0.15	0.02	0.02	0.17	0.08	0.37	0.15	0.22	0.26
Q9b3					1.00	0.39	0.27	0.23	0.22	0.09	0.20	0.15	0.37
Q9b4						1.00	0.44	0.28	0.27	0.02	0.22	0.14	0.20
Q9b5							1.00	0.48	0.30	0.09	0.28	0.22	0.21
Q9b6								1.00	0.39	0.25	0.37	0.37	0.22
Q9b7									1.00	0.19	0.25	0.23	0.18
Q9b8										1.00	0.29	0.37	0.33
Q9b9											1.00	0.63	0.30
Q9b10												1.00	0.36
Q9b11													1.00

5. Methodology and Results

Here we re-examine the four requirements for supply chains to have contributed to global diffusion of ISO 9000, and discuss our methodology for testing each of these and the results obtained. For brevity, we use "area" to refer to the specific countries in which the survey was conducted or to the geographic region to which the respondent firms export.

5.1 Requirement 1: Heterogeneity Across Areas of Timing of Adoption

To assess the extent of adoption in a country at any given time, we would ideally compare the number of certifications to the maximum number of potential certifications, e.g., the number of establishments, but the only global data that come close to this, published by UNIDO (2003), have notable shortcomings for our purposes. To overcome these limitations, we use four different methods to assess R1: ranking of certification levels deflated by GDP, population, number of establishments, and a ranking by relative growth rates derived from the Bass diffusion model. Our data do not allow meaningful statistical tests of the relative timing of adoption, so our discussion of the first requirement will necessarily be more qualitative than that of the other three. Two other potential methods also do not apply here. First, the one-sided Kolmogorov–Smirnov test or the Wilcoxon–Mann–Whitney test, frequently used for testing first-order stochastic dominance, would only be applicable if the maximum number of potential certifications was known. Second, comparing mean year of adoption across countries does not work as, for all countries, the bulk of certifications takes place in the late 1990s. The mean year of certification for all countries based on the sample was between 1,995.2 and 1,998.6 with seven out of nine countries in 1995–1997 range, too close to be able to distinguish statistically. Simple numerical experiments based on the Bass diffusion model also confirm that the difference between average year of certification in two countries is much smaller than the difference between year of first certification in those countries.

The first two approaches to ranking regions and countries by timing of adoption involve deflating certification counts using GDP and population respectively, both imperfect but not unreasonable measures for this purpose. GDP has been used as a deflator in Guler et al. (2002) and Corbett and Kirsch (2001); however, this overcorrects in rich countries as the US and

Japan relative to developing countries. Deflating by population does the opposite, overcorrecting in large developing countries such as China and India relative to small rich countries. Figure 2 shows the diffusion by region, starting in January 1993 with the earliest existing data. At that time, ISO 9000 had existed for 7 years and had been adopted by some 27,000 sites in 48 countries (ISO, 2004). Visual inspection of Fig. 2 shows that ISO 9000 first spread within Australia/New Zealand and Europe, with other areas following later, regardless of whether one deflates by GDP or population. Table 6 shows the rankings for each region based on each of the four methods. (Regions include all countries within the usual geographical boundaries, not just those included in the survey.)

Similarly, Fig. 3 indicates that Australia/New Zealand, Canada, France, Hong Kong and Sweden were early-adopting countries, with Japan, Korea, Taiwan and the US following later. (We consistently refer to "later-adopting" rather than "late-adopting" as the survey does not include truly late-adopting countries.) This is again consistent with expectations: the early-adopting countries are in Europe or have close ties to the UK, where the BS 5750 standard, a partial predecessor of ISO 9000, was already well-established when ISO 9000 was introduced (Guler et al., 2002). Table 7 shows the rankings for the countries based on each of the four methods.

Table 6. Ranking of regions according to earliness of ISO 9000 adoption, according to various criteria; regions that are substantially earlier than the rest are indicated in bold

	cum. certifications Jan 1993	new certifications Sep 1993	GDP 1993 (in million US$)	population 1993 (in million inhabitants)	establishments 1993	cum. cert. / GDP Jan 1993	rank	cum. cert. / pop. Jan 1993	rank	cum. cert. / est. Jan 1993	rank	new cert. Sep 1993 / cum. cert. Jan 1993	rank
Australia / NZ	1862	1322	328,649	21	60665	5.666	1	88.318	1	3.07%	1	0.71	3
Europe	23092	14682	8,600,484	630	1265815	2.685	2	36.612	2	1.82%	3	0.64	2
Africa	825	184	404,795	573	34440	2.038	3	1.440	4	2.40%	2	0.22	1
North America	1201	1412	7,886,889	375	442401	0.152	5	3.195	3	0.27%	4	1.18	4
Japan	165	269	3,947,971	123	396632	0.042	6	1.337	5	0.04%	5	1.63	6
Asia w/o Japan	644	773	3,222,152	3	1755483	0.200	4	0.202	6	0.04%	6	1.20	5
South America	27	113	1,402,918	362	229123	0.019	7	0.074	7	0.01%	7	4.19	7

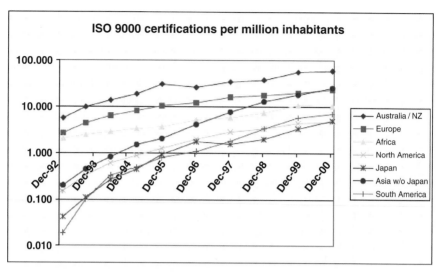

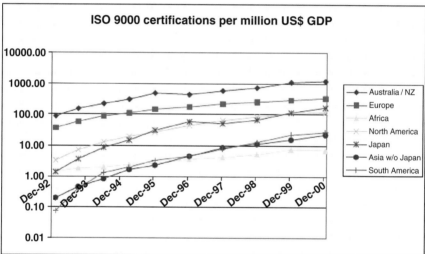

Figure 2. ISO 9000 certificates by region. The *upper* chart is certifications per million inhabitants, the *lower* is certifications per million US$ GDP. Both charts are shown on a logarithmic scale

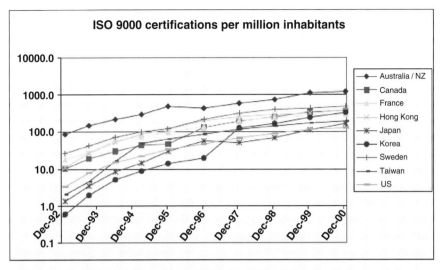

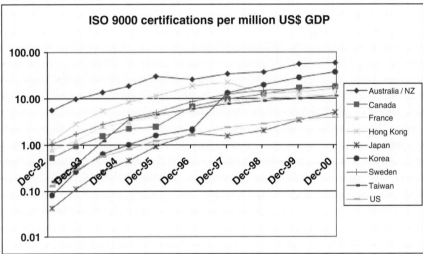

Figure 3. ISO 9000 certificates by country. The *upper* chart is certifications per million inhabitants, the *lower* is certifications per million US$ GDP. Both charts are shown on a logarithmic scale

Table 7. Ranking of countries according to earliness of ISO 9000 adoption, according to various criteria; countries that are substantially earlier than the rest are indicated in bold

	cum. certifications Jan 1993	new certifications Sep 1993	GDP 1993 (in million US$)	population 1993 (in million inhabitants)	establishments 1993	cum. cert. / GDP Jan 1993	rank	cum. cert. / pop. Jan 1993	rank	cum. cert. / est. Jan 1993	rank	new cert. Sep 1993 / cum. cert. Jan 1993	rank
Australia / NZ	1862	1322	328,649	21	60665	5.666	1	88.318	1	3.07%	2	0.71	3
Canada	292	238	557,079	29	31630	0.524	5	10.245	5	0.92%	4	0.82	4
France	1049	537	1,322,384	57	22438	0.793	4	18.377	3	4.68%	1	0.51	1
Hong Kong	69	92	57,266	6	36847	1.205	2	11.226	4	0.19%	6	1.33	7
Japan	165	269	3,947,971	123	396632	0.042	9	1.337	8	0.04%	7	1.63	8
Korea	27	60	335,101	44	88864	0.081	8	0.615	9	0.03%	8	2.22	9
Sweden	229	136	219,625	9	8153	1.043	3	26.379	2	2.81%	3	0.59	2
Taiwan	43	53	267,539	21	n/a	0.161	6	2.056	7	n/a	n/a	1.23	5
US	893	1166	6,960,440	261	403614	0.128	7	3.420	6	0.22%	5	1.31	6

Turning to our third method for assessing the first requirement, we use the UNIDO (2003) data on number of establishments by country. These data do not distinguish establishments by size or sector, so they are not an accurate estimate of the total market potential for ISO 9000 certification in a country; moreover, no data are available for Taiwan. Tables 6 and 7 provide the rankings of regions and countries based on certifications as of January 1993 divided by estimated number of establishments in 1993. (Using certification and establishment data for 1994 yields an almost identical ranking, and does not change the classification into early- and later-adopting areas.) Here, the three early-adopting regions are Australia/New Zealand, Africa, and Europe. With respect to the ranking of countries, Table 7 shows that the only change compared to the earlier ranking is that Hong Kong now moves into the later-adopting category.

For the fourth perspective, consider that the Bass model implies that the ratio of new adoptions ΔN_t to cumulative past adoptions N_{t-1} must always be declining over time (see Appendix A). Tables 6 and 7 rank the regions and countries by their Δ values for September 1993. Africa, Europe and Australia/New Zealand have the lowest scores, implying that they were furthest along the diffusion curve in 1993. In the country ranking, Hong Kong would again be a later adopter, but otherwise the ranking is consistent with the earlier approaches. $\Delta N_t / N_{t-1}$

In conclusion, widespread adoption occurred first in Europe, though Australia/New Zealand and, by some criteria, Africa also can be considered early-adopting regions. North America, South America, Japan, and Asia without Japan are the later-adopting regions. This is largely consistent with the view held widely among scholars and practitioners of ISO 9000 that diffusion of ISO 9000 started in Europe (e.g., Anderson et al., 1999); Australia/New Zealand and certainly Africa, despite being early adopters by our relative measures, may not have sufficiently high absolute numbers of adoptions or be sufficiently large regions to actually drive adoption elsewhere, as we see later. Similarly, we can classify Australia/New Zealand, Canada, France and Sweden as the early-adopting countries, with Japan, Korea, Taiwan and the US being the later-adopting countries; we cannot unambiguously classify Hong Kong. In summary, Figs. 2 and 3 and Tables 6 and 7 show that the first requirement is met: there exists considerable heterogeneity between areas with respect to timing of widespread adoption.

5.2 Requirement 2: Firms in Early-Adopting Regions Exert More Pressure

Having classified Africa, Australia/New Zealand and Europe as early-adopting regions and the rest as later-adopting, the second requirement can be reformulated as follows:

Hypothesis R2: The proportion of customers in Africa, Australia/New Zealand and Europe requiring ISO 9000 certification is higher than that in other regions.

To test this, we use questions 6 and 7 in the survey. We computed the mean for question 7a, the proportion of customers in region A requiring ISO 9000 certification. Table 8 shows the results by country and for the entire sample, where "1" means no customers in that region require certification and "5" means that all customers do. In the global sample and in all countries except Hong Kong, firms report that the proportion of customers in Europe requiring certification is higher than that in other regions. For instance, in the global sample, the mean for Europe is 3.37, so the proportion of customers requiring certification lies between "some" (3) and "most" (4). The scores for Australia/New Zealand and especially Africa are much lower. appropriate test with paired observations is the one-sided Wilcoxon signed-rank test (Miller et al., 1999), which involves computing all pairwise differences between regions, and calculating a test statistic based on the ranking of the absolute differences and their sign. We used PROC UNIVARIATE in SAS for the latter part of this procedure.

Table 8. Hypothesis R2: proportion of customers in different regions requiring certification, by country (survey question 7a)[30]

respondent countries:	customer region						
	Europe	Japan	North America	South America	Asia w/o Japan	Australia / NZ	Africa
Australia / New Zealand	2.57 (1.54)	2.29 (1.46)	2.36 (1.53)	1.75 (1.23)	2.38 (1.32)	2.92 (1.22)	1.59 (1.02)
Canada	3.54 (1.38)	2.74 (1.52)	3.49 (1.23)	2.23 (1.44)	2.34 (1.27)	2.40 (1.45)	1.62 (0.99)
France	3.04 (1.24)	2.29 (1.57)	2.45 (1.52)	2.07 (1.33)	2.31 (1.40)	n/a	n/a
Hong Kong	3.22 (1.13)	2.67 (1.33)	2.40 (1.22)	2.31 (1.20)	3.36 (1.01)	2.67 (1.23)	2.14 (1.23)
Japan	3.42 (1.05)	3.19 (0.86)	3.07 (1.04)	2.44 (0.92)	2.72 (0.94)	n/a	n/a
Korea	3.92 (1.05)	3.66 (1.15)	3.72 (1.20)	3.03 (1.14)	3.53 (1.08)	3.23 (1.17)	2.55 (1.03)
Sweden	3.59 (1.00)	2.85 (1.49)	2.81 (1.50)	2.30 (1.39)	2.27 (1.26)	2.63 (1.45)	1.82 (1.15)
Taiwan	3.81 (1.25)	3.61 (1.32)	3.58 (1.27)	3.02 (1.31)	3.24 (1.13)	3.23 (1.35)	2.11 (1.14)
US	3.38 (1.35)	2.71 (1.41)	3.20 (1.20)	2.43 (1.32)	2.49 (1.30)	2.44 (1.32)	1.93 (1.18)
Global	3.37 (1.24)	3.08 (1.11)	3.12 (1.24)	2.43 (1.21)	2.76 (1.17)	2.77 (1.32)	1.94 (1.15)

For each pair of regions, Table 9 shows the mean difference in proportion of customers from those regions requiring certification, using the pooled data across all countries. For instance, respondents reported a 0.28 higher score for proportion of customers in Europe than in North America requiring certification, and this is significant with a *p*-value of 0.000. This difference of 0.28 between Europe and North America is close to but not equal to the difference between the European mean of 3.37 and the North American mean of 3.12 in Table 8, because the means in Table 8 are computed over all firms that responded to the question pertaining to that one region (e.g., Europe) while the means in Table 9 are taken over all firms that responded to the questions pertaining to both regions concerned (e.g., Europe and North

[30] The first number in each cell indicates the mean proportion of customers from the column region requiring respondents in the row country to seek certification; the second number (between brackets) is the standard deviation. Tables 9 and 10 show results of tests for differences between regions. The French and Japanese surveys omitted Africa and Australia/New Zealand from the list of regions.

America). From Table 9 we see that Europe has the highest proportion of customers requiring ISO 9000 certification, followed by North America and Japan, then Asia, Australia/New Zealand, South America and Africa.

Table 9. Hypothesis R2: mean difference between proportion of customers in different regions requiring ISO 9000 certification, using pooled data (survey question 7a)[31]

	Europe	Japan	North America	South America	Asia w/o Japan	Australia / New Zealand	Africa
Europe		*0.30* 0.000 32.2%	*0.28* 0.000 31.0%	*0.94* 0.000 57.2%	*0.65* 0.000 49.1%	*0.69* 0.000 43.9%	*1.29* 0.000 60.6%
Japan	*-0.30* 0.000 13.3%		*0.01* 0.719 25.1%	*0.64* 0.000 46.6%	*0.39* 0.000 39.5%	*0.14* 0.000 24.4%	*0.78* 0.000 40.6%
North America	*-0.28* 0.000 10.2%	*-0.01* 0.719 22.1%		*0.63* 0.000 45.0%	*0.33* 0.000 35.4%	*0.42* 0.000 39.4%	*1.06* 0.000 58.0%
South America	*-0.94* 0.000 2.8%	*-0.64* 0.000 7.0%	*-0.63* 0.000 3.6%		*-0.30* 0.000 10.2%	*-0.22* 0.000 13.2%	*0.38* 0.000 26.6%
Asia w/o Japan	*-0.65* 0.000 6.1%	*-0.39* 0.000 9.2%	*-0.33* 0.000 14.4%	*0.30* 0.000 31.0%		*-0.06* 0.085 20.7%	*0.65* 0.000 41.4%
Australia / New Zealand	*-0.69* 0.000 5.3%	*-0.14* 0.000 13.8%	*-0.42* 0.000 13.3%	*0.22* 0.000 26.4%	*0.06* 0.085 23.1%		*0.59* 0.000 37.5%
Africa	*-1.29* 0.000 1.0%	*-0.78* 0.000 3.3%	*-1.06* 0.000 3.5%	*-0.38* 0.000 3.6%	*-0.65* 0.000 2.7%	*-0.59* 0.000 2.0%	

Given the large sample size, finding highly significant differences between regions is not surprising. More insightful is to compare the percentage of respondents experiencing higher pressure from region A than region B with the percentage experiencing higher pressure from region B

---

[31] The first number in each cell indicates the mean difference (row minus column); the second gives the p-value for the one-sided Wilcoxon signed-rank test; the third indicates the percentage of respondents reporting a strictly higher score for the row region than for the column region. E.g., firms reported more European customers requiring ISO 9000 certification; the mean difference between Europe and North America was 0.28. 31.0% of those respondents reported a strictly higher score for Europe, 10.2% reported a strictly higher score for North America. The number of observations for each comparison depends on the number of firms responding about exports to both regions, but was always between 490 and 2518, hence the high degrees of significance.

than region A. For instance, 31.0% of respondents report a strictly greater proportion of customers in Europe than in North America requiring ISO 9000 certification, and only 10.2% report the opposite. The same pattern is largely replicated within each country, though due to the smaller sample sizes within countries, some differences are no longer significant. Table 10 shows, for each country, which differences are positive and significant at the 1 and 5% levels; Europe clearly still leads.

Table 10. Results for Hypothesis R2: significant differences between proportion of customers in different regions requiring ISO 9000 certification (survey question 7a) by country[32]

	Europe	Japan	North America	South America	Asia w/o Japan	Australia / New Zealand	Africa
Europe		*C,F,S,U* *A,J,T*	*F,J,S,T,U* *A,H,K*	*C,F,H,J,K,S,T,U* *A*	*A,C,F,J,S,T,U*	*C,K,S,T,U* *H*	*C,H,K,S,T,U*
Japan			*J*	*C,F,J,K,S,T,U*	*C,J,S,T,U*	*K,T,U* *C*	*C,K,S,T,U*
North America		*C,U*		*C,F,J,K,S,T,U*	*C,J,S,U* *T*	*C,K,T,U*	*C,K,S,T,U* *H*
South America							*K,T,U* *C*
Asia w/o Japan		*H*	*H*	*A,F,H,J,K,U* *C,S,T*		*K* *H,U*	*A,C,H,K,S,T,U*
Australia / NZ		*A*	*A*	*A,S* *C,K*	*A* *S*		*A,C,K,S,T,U* *H*
Africa							

Overall, we find strong support for the hypothesis that a higher proportion of customers from Europe require certification than from other regions, but no support for the corresponding hypothesis for Africa and only minimal support for Australia/New Zealand, whose firms do exert much more pressure than those in Africa or South America but less than those in Europe, Japan and North America. Hence, we drop Africa as a potential driver of ISO 9000 certification from here on.

---

[32] If respondents in country i report a significantly higher proportion of customers in region A requiring certification than in region B, that country's letter is entered in row A, column B. This uses the one-sided Wilcoxon signed-rank test at the 5% (italics) and 1% (bold italics) significance levels. For instance, the entries "K,T,U" and "C" in row Japan and column Australia/New Zealand indicate that among Korean, Taiwanese and US respondents the average response for Japan was significantly higher than that for Australia/New Zealand at the 1% level and for Canadian respondents at the 5% level. In many cases, a positive difference was not significant as too few respondents in country i had answered the questions about exports to both region A and B.

5.3 Requirement 3: In Later-Adopting Areas, Firms with Higher Exports to Early-Adopting Areas Adopt Earlier

Our third and fourth requirements focus on countries in which diffusion started relatively late, i.e., Japan, Korea, Taiwan and the US. Including the appropriate control variables, requirement 3 becomes:

Hypothesis R3: in Japan, Korea, Taiwan and the US, firms with higher exports to Australia/New Zealand or to Europe certified earlier, after correcting for size and other exports.

Our dependent variable here is ISO9Year, the year of original certification for each respondent. We get "Exports to Other Regions" by computing the mean of the responses to question 6 over all regions except Australia/New Zealand and Europe. Controlling for size is slightly more involved: questions 1 and 2 asked about employees and sales respectively, but many respondents only answered one of the two questions, so that any analysis using the original scores for questions 1 or 2 would miss many responses. We define a company as "large" (Large = 1) if it has more than 1,000 employees worldwide (question 1b) or falls within the top quartile (within its country) with respect to world sales (question 2b). This new dummy variable exhibits high Spearman rank correlation with the two original size variables, generally in the 0.75–0.85 range and always greater than 0.60, so it is an acceptable proxy for size. Setting the cutoff point at 5,000 employees worldwide gives similar results, as does the analysis based on either of the original variables but using less observations.

Hypothesis R3 then predicts that ISO9Year is negatively associated with Exports to Australia/New Zealand and with Exports to Europe when correcting for size and exports to all other regions. In testing this, we face two challenges. We need to allow for the possibility that there exist interactions between firm size and each of the export variables, which would lead to unequal slopes for smaller and larger firms. We also need to incorporate the fact that ISO9Year is truncated above (Maddala, 1983, Sect. 6.9), as the data do not include certifications that occurred after the survey was administered. No established estimation procedure takes both of these issues into account, so we deal with them separately. We tested for interactions using analysis of covariance (ANCOVA, see Hair et al., 1998), using PROC GLM in SAS, but only found a significant interaction effect in Japan, due to that country's large sample. This justifies focusing on the truncation problem while ignoring interaction effects. Therefore we use the maximum likelihood estimator for truncation regression (Maddala, 1983) as implemented in PROC QLIM in SAS, setting the upper truncation for each country equal to the year of its most recent certification in the survey (Table 11).

Table 11. Results for Hypothesis R3: explaining year of adoption[33]

country	N	intercept		Large		Exports to Europe		Exports to Australia / NZ		Exports to Other Regions	
		estimate p-value		estimate p-value		estimate p-value		estimate p-value		estimate p-value	
Australia/NZ	576	1997 <.0001	***	-1.2717 <.0001	***	-0.0242 0.8366		-0.1663 0.1692		-0.1065 0.5375	
Canada	181	1998 <.0001	***	-1.8577 0.0011	***	0.0116 0.9598		-0.6923 0.0688	*	0.0414 0.9411	
(France)	424	2013 <.0001	***	-2.2694 0.0025		-1.4717 0.0060	***	n/a		-3.1341 <.0001	***
Hong Kong	112	2001 <.0001	***	-4.8316 <.0001	***	0.8993 0.1483		0.0858 0.8877		-1.0853 0.3556	
Japan	2164	2004 <.0001	***	-2.1383 <.0001	***	-0.5800 <.0001	***	n/a		-1.2455 <.0001	***
Korea	111	1997 <.0001	***	-1.5909 0.0029	***	-0.6196 0.0061	***	0.3642 0.1996		0.1894 0.6923	
Sweden	133	2001 <.0001	***	-2.7881 0.0002	***	-0.4915 0.4807		0.7579 0.0972		-0.9727 0.0541	*
Taiwan	434	1997 <.0001	***	-0.6611 0.0024	***	-0.3177 0.0030	***	0.0313 0.8022		-0.2603 0.2150	
US	891	1998 <.0001	***	-0.9226 <.0001	***	-0.1725 0.0229	**	0.1489 0.1487		-0.4689 0.0057	***

Table 12. Results for Hypothesis R3: explaining year of adoption[34]

country	parameter estimate and significance level for Exports to:													
	Europe		Japan		North America		South America		Asia w/o Japan		Australia / New Zealand		Africa	
Australia/NZ	-0.04		0.24	**	-0.14		-0.02		-0.18	**	-0.14		0.15	
Canada	0.03		0.03		1.26	***	0.04		-0.03		-0.63	*	-0.13	
(France)	-1.47	***	-0.33		-0.61	*	-0.19		-0.81	**	n/a		n/a	
Hong Kong	0.29		-0.39		-0.02		0.97		-1.58	***	-0.01		-0.07	
Japan	-0.50	***	0.08		-0.35	***	0.18	*	-0.40	***	n/a		n/a	
Korea	-0.62	**	-0.02		0.00		0.19		-0.04		0.27		0.19	
Sweden	-0.53		-0.07		0.13		-0.40		-0.43		0.83	*	-0.39	
Taiwan	-0.35	***	-0.02		-0.12		0.14		-0.15		0.02		-0.19	
US	-0.15	**	0.10		0.16		-0.04		-0.34	***	0.15		-0.10	

---

[33] These results are maximum likelihood estimates for the truncated regression by country. Significance at the 1% level is shown by ***, at the 5% level by **, at the 10% level by *. The figures for France are not reliable, as some respondents may have reported the most recent recertification rather than the year of original certification, and firms must recertify to ISO 9000 every three years. However, as hypotheses R3 and R4 focus on later-adopting countries, the results for (early-adopting) France do not affect our conclusion in support of hypotheses R3 and R4. Note also that the classification of France as an early-adopting country was based on national certification data, not on the survey data. The data for France are shown only for completeness.

[34] These results are maximum likelihood estimates for the truncated regression by country. Significance at the 1% level is shown by ***, at the 5% level by **, at the 10% level by *. The figures for France are not reliable and are shown only for completeness, see note in Table 11.

Table 12 shows that Exports to Europe is significant at the 1% level in Japan, Korea, and Taiwan, and at the 5% level in the US. This means that Exports to Europe are a significant factor in explaining early adoption in all four later-adopting countries, after correcting for size and exports to all other regions. Exports to Australia/New Zealand have little or no effect on adoption timing in most countries. There is of course some correlation between Exports to Australia/New Zealand, Exports to Europe and Exports to Other Regions (the correlation coefficients vary between 0.06 and 0.45); the consequence of such moderate multicollinearity is usually to inflate the standard errors and hence reduce the significance of the parameter estimate. Therefore, this correlation does not weaken the finding that Exports to Europe are strongly significant. To avoid this correlation, we can combine all the "Exports to ..." variables into one single variable, Global Exports (results not reported); this new variable is significant at the 1% level for Japan, Taiwan, and the US but not significant for Korea. To examine the magnitude of each regions' impact separately, at the expense of exacerbating multicollinearity, we repeat the analysis with separate "Export to ..." variables for each region, shown in Table 12. Exports to Europe are again significant in Japan, Korea, Taiwan and the US. Exports to North America drive earlier certification in Japan but later certification in Canada; exports to Asia cause earlier certification in Australia/New Zealand, Hong Kong, Japan and the US. Exports to Africa and Exports to Australia/New Zealand have little or no effect. In most cases, the parameter estimate for Exports to Europe is the largest (in absolute terms), indicating that exports to Europe is a stronger driver of early certification than exports to any other individual region. Overall, this provides fairly strong support that the third requirement is met with respect to Europe but not Australia/New Zealand.

5.4 Requirement 4: In Later-Adopting Countries, Firms that Adopt Later are Less Motivated by Export-Related Factors

To examine this, we need to define "early" and "later" adopting firms. Firms with certification year in the first tertile within each country are defined as "early" adopters, those with certification year in the third tertile as "later" adopters, leaving out the middle third. (Using other classifications, based on quartiles or median, and with or without omitted middle groups, gave highly similar results.)

Question 9b asked about motivations for certification. The motivation that explicitly addresses exports is "avoid potential export barrier." The hypothesis corresponding to our fourth requirement is then:

Hypothesis R4: in Japan, Korea, Taiwan and the US, firms that adopt later are less motivated by export protection than are earlier adopters.

A bias would occur if early adopters reported higher scores for all motivations; therefore, we normalize each motivation score by dividing it by the sum over all 11 motivation factors. Let $MOT(k)_i$ be respondent i's score for motivation k; the relative score for motivation k is $RELMOT(k)_i := MOT(k)_i / \sum_{j=1}^{11} MOT(K)_j$, The normalized variable for the export motivation is RELMOTEXP. (Using the raw scores does give similar results.)

The test is now whether, in the four later-adopting countries, firms classified as early adopters have higher scores for RELMOTEXP than later adopters. The "customer pressure" motivation factor (RELMOTCUST) does not distinguish between domestic and foreign customers, so our set of requirements does not predict how the importance of this factor differs between early and late adopters.

The normalized motivation scores are not normally distributed, so we apply nonparametric analysis of variance (ANOVA), using the one-sided Wilcoxon two-sample test (Miller et al., 1999), which is based on a test statistic calculated using the ranks of each observation within their respective group. We used PROC NPAR1WAY in SAS. (Note that multivariate ANOVA, or MANOVA, is not appropriate here as we are interested in the effect of early adoption on individual motivation factors, not on all motivations jointly.) Table 13 shows significant differences (at the 1% level) in RELMOTEXP between early and later adopters in the expected direction in all four later-adopting countries. (The hypothesis does not preclude finding the same pattern in early-adopting countries, as we do in Australia/New Zealand.) This is not a mere sample size effect: in Japan, for instance, almost any difference would be significant due to the large sample, but the magnitude of the difference is also larger than for most other motivations. The descriptive statistics in Table 5 serve as a reminder that other factors do contribute to explaining adoption, as most other motivations display higher average scores than export protection does. However, the variance of the scores for export protection is the highest, and it is the only motivation that strongly distinguishes early and later adopters in the four later-adopting countries, hence providing support for hypothesis R4.

Table 13. Results for Hypothesis R4: differences between early and later adopters in relative motivation factors[35]

country	relative motivation factors: see question 9b in the Appendix										
	cost reductions	quality improvements	marketing advantage	customer pressure / customer demands	many competitors were already ISO 9000 certified	benefits experienced by other certified companies	avoid potential export barrier	capturing workers' knowledge	relations with authorities	relations with communities	corporate image
	1	2	3	4	5	6	7	8	9	10	11
Australia / NZ		---		+++			+++		--	---	---
Canada									+++	++	
(France)	-			++	++		++		--	--	
Hong Kong		--		++	++		-	--			
Japan	---	---		+++		++	+++	---	---	---	---
Korea			-				+++	-		---	--
Sweden			+			---					
Taiwan	-	-	+	+	-		+++				-
US	-	--				--	+++	-			

6.   Discussion, Conclusions and Future Research

Our findings can be summarized as follows. Diffusion of ISO 9000 started (primarily) in Europe in large numbers. From Europe, it spread to other countries, as customers in Europe pressured foreign suppliers to seek ISO 9000 certification. Those suppliers in other countries then sought certification as protection against the perceived threat of ISO 9000 becoming an export barrier. After exporting firms in other countries had adopted the standard, domestic diffusion by traditional mechanisms picked up, and other firms in those countries started adopting ISO 9000 too, resulting in over 560,000 certifications in 152 countries by 2003. The findings in this paper suggest that at least part of those certifications can be explained by pressure exerted by downstream customers through global supply chains on upstream

---

[35] These results are based on nonparametric ANOVA by country, using the t-approximation for the one-sided Wilcoxon two-sample test of differences of means. A "+" indicates that early adopters have a higher score, "−"indicates a lower score; +++ indicates significance at the 1% level, ++ at 5%, and + at 10%. "Early" and "later" adopters are defined as the first and third tertile respectively, the middle tertile is omitted. These results for France are unreliable and shown only for completeness; see note in Table 11.

firms in other countries. This implies that firms exporting goods or services into a region may also simultaneously import management practices from that region back into their home country.

Firms in Africa and Australia/New Zealand, the other regions that we classify as early-adopting by our relative measures, did not exert as much pressure on suppliers elsewhere to seek certification, and appear to have contributed less or not at all to the global diffusion of ISO 9000 despite being early adopters themselves. One might speculate that this is due to the fact that individually and collectively, most firms in Africa and in Australia/New Zealand are smaller and less powerful than those in Europe, but our data do not allow us to verify that.

This paper contributes to the diffusion literature, by explicitly examining the role of supply chains in explaining global diffusion of management practices, and by formulating precise requirements for claiming that supply chains contribute to global diffusion. This paper also contributes to the literature on ISO 9000, by testing hypotheses derived from these requirements and showing that ISO 9000 did diffuse worldwide through global supply chains.

This study has several inevitable limitations. It does not include truly late-adopting countries; attempts were made to administer the survey in Argentina, Mexico and South–East Asia, but these were largely unsuccessful. The survey, administered around 2000, is an imperfect way of obtaining information about motivations for certifications that occurred long before; however, the bulk of all certifications, including those in early-adopting countries, occurred in the mid- to late-1990s, which partially mitigates this problem. Despite these and other limitations which are inherent in survey research, two factors do lend additional credibility to the results. First is their face validity: the picture that emerges from this analysis is consistent with anecdotal views of global diffusion of ISO 9000. Second is their robustness: each of the tests reported here was performed in several different ways, without leading to substantially different results.

This study has only examined ISO 9000, so we cannot conclude that other management practices or standards diffuse through global supply chains too. Documenting similar findings for ISO 14000 and other standards and practices is necessary to assess the generalizability of our findings beyond ISO 9000. However, this author's original interest was in ISO 14000, the environmental management systems standard introduced in 1996 and adopted by over 66,000 sites in 113 countries by December 2003 (ISO, 2004). The design of the ISO 14000 standards is analogous to that of ISO 9000, even though its scope is different. If one can demonstrate that an environmental management standard as ISO 14000 diffuses through global supply chains, that would imply that the notion of "greening the supply chain" is indeed a feasible non-governmental market-based mechanism for spreading environmental practices to other countries. The current study cannot answer that question for ISO 14000, which is left for future research, but this paper will hopefully contribute indirectly to that debate by showing that this mechanism did occur for its predecessor, ISO 9000.

Appendix A: Relating Our Frame Work to Bass Diffusion Model

The Bass (1969) diffusion model relates ΔN_t, the number of new adoptions in period t, to cumulative past adoptions through the following relationship: N_{t-1}

$$\Delta N_t := N_t - N_{t-1} = \left(a + b \frac{N_{t-1}}{M} \right)(M - N_{t-1}) \qquad (1)$$

where M is the saturation level, a constant diffusion rate and b the coefficient of the imitation effect. This model gives rise to the classic S-shaped diffusion curve. Note that $\dfrac{\Delta N_t}{N_{t-1}} = \left(\dfrac{a}{N_{t-1}} + \dfrac{b}{M} \right)(M - N_{t-1})$ always decreasing in t, the observation underlying our fourth method for classifying regions and countries into early and later adopters.

To model a global diffusion process, one could attach a country index i to all variables in (1), estimate the resulting diffusion curves, and compare the resulting coefficients a_i, b_i, and M_i across countries. This is the approach taken in Lücke (1996). To explicitly capture diffusion across K countries, we can use the following adaptation, which is also the basis for Albuquerque et al. (2005):

$$\Delta N_{it} := N_{it} - N_{i,t-1} = \left(a_i + \sum_{j=1}^{K} b_{ij} \frac{N_{j,t-1}}{M_j} \right)(M_i - N_{i,t-1}) \qquad (2)$$

New adoptions in country i in period t depend on the constant diffusion rate a_i and the domestic imitation effect $b_{ii} N_{i,t-1} / M_i$, but also on $\sum_{i \neq j} b_{ij} N_{j,t-1} / M_j$, where b_{ij} denotes the influence of country i on country j. This is analogous to the multi-product diffusion model in Bayus et al. (2000).

Expression (2) implies that the effect of country j on country i becomes stronger with the penetration level in country j, resembling our second requirement. In later-adopting countries, in the early years, the domestic imitation effect $b_{ii} N_{i,t-1} / M_i$ is small compared to the cross-country effect $\sum_{r=1}^{R} b_{ir} N_{r,t-1} / M_r$ as, by definition, $N_{i,t-1} / M_i$ is small relative to at least some of the other countries' $N_{r,t-1} / M$. Over time $N_{i,t-1} / M_i$ may catch up with $N_{r,t-1} / M$, so that the domestic imitation effect gains strength relative to the cross-country effect. Therefore, in countries in which diffusion started late, the cross-country effect should play a larger role for early-adopting firms than for later-adopting firms. This is consistent with our third and fourth requirements, where exports are the cross-country effect.

Appendix B: Excerpts from Questionnaire Used

1. How many employees work at your company, at the same facility as you? And how many work at your company, counting all locations world wide?

1a. At your facility:	1: 1 – 19	2: 20 – 99	3: 100 - 499	4: 500 - 999	5: 1000 – 4,999	6: 5,000 - 24,999	7: 25,000 or more
1b. Worldwide:	1: 1 – 19	2: 20 – 99	3: 100 - 499	4: 500 - 999	5: 1000 – 4,999	6: 5,000 - 24,999	7: 25,000 or more

2. Please estimate the total annual sales generated by your facility and the total global annual sales of your company, in US$.

Total annual sales at your facility (US$):	
Total global, annual sales of your entire company (US$):	

3. Please tell us where your *immediate* customers are located, by indicating how important sales in each of the following geographic regions are for your company.

	not at all important	not important	moderately important	important	very important
North America	1	2	3	4	5
South America	1	2	3	4	5
Europe	1	2	3	4	5
Africa	1	2	3	4	5
Australia / New Zealand	1	2	3	4	5
Japan	1	2	3	4	5
Asia (except Japan)	1	2	3	4	5

4a. Please indicate how many customers in each of the following regions require ISO 9000 certification or are expected to do so soon.

	no customers in this region	proportion of customers requiring ISO 9000 (now or in near future)				
		none	few	some	most	all
North America	0	1	2	3	4	5
South America	0	1	2	3	4	5
Europe	0	1	2	3	4	5
Africa	0	1	2	3	4	5
Australia / New Zealand	0	1	2	3	4	5
Japan	0	1	2	3	4	5
Asia (except Japan)	0	1	2	3	4	5

4b. Please indicate how important each of the following reasons was for seeking, maintaining, or seriously considering ISO 9000 certification. (If you do not have ISO 9000 certification and are not considering it, please do not answer this question.)

	not important at all	not important	somewhat important	important	extremely important
cost reductions	1	2	3	4	5
quality improvements	1	2	3	4	5
marketing advantage	1	2	3	4	5
customer pressure / customer demands	1	2	3	4	5
many competitors were already ISO 9000 certified	1	2	3	4	5
benefits experienced by other certified companies	1	2	3	4	5
avoid potential export barrier	1	2	3	4	5
capturing workers' knowledge	1	2	3	4	5
relations with authorities	1	2	3	4	5
relations with communities	1	2	3	4	5
corporate image	1	2	3	4	5

Acknowledgements

The author is deeply grateful to the partners in this global survey for administering the survey in their respective countries: Mile Terziovski, Rob Klassen, Olivier Aptel, TY Lee, Kazuhiro Asakawa, Phares Parayno, Carlos Romero, Hosun Rhim, and Jens Dahlgaard. I am particularly grateful to Jack Pan for his assistance in developing the survey and administering it in Taiwan, and to David Kirsch for many discussions on related subjects and for suggesting the "golden dollar" incentive. Don Morrison gave insightful suggestions on interpretation of some of the statistical results. I am also grateful to Anastasia Luca and Maria J. Montes for their research assistance and to Theresa Jones and Susana Medina for their assistance in administering the survey, and to WorldPreferred for providing the database of US certifications. Thanks are also due to the UCLA Pacific Rim Research Program, CIBER at the UCLA Anderson School of Management, to ISO, and the James Peters Research Fellowship for their financial support. Finally, the Senior Editor's and reviewers' constructive comments were very helpful in revising this manuscript.

References

Abrahamson, E., 1991, Managerial Fads and Fashions: The Diffusion and Rejection of Innovations., *Academy of Management Review*, 16 (3): 586–612.

Albuquerque, P., Bronnenberg, B., and Corbett, C.J., 2005, A Spatio-Temporal Analysis of the Global Diffusion of ISO 9000 and ISO 14000 Certification, *Working paper*, UCLA Anderson School of Management.

Anderson, S.W., Daly, J.D., and Johnson, M.F., 1999, Why Firms Seek ISO 9000 Certification: Regulatory Compliance or Competitive Advantage? *Production and Operations Management* 8 (1): 28–43.

Bass, F.M., 1969, A New Product Growth Model for Consumer Durables, *Management Science* 15 (5): 215–227.

Bayus, B.L., Kim, N., and Shocker, A.D., 2000, Growth Models for Multiproduct Interactions: Current Status and New Directions. Chapter 7 in Mahajan, V., E. Muller and Y. Wind (eds.), *New-Product Diffusion Models*, Kluwer Academic Publishers, Boston.

Brown, A., van der Wiele, T., and Loughton, K., 1998, Smaller Enterprises' Experiences with ISO 9000. *International Journal of Quality & Reliability Management* 15 (3): 273–285.

Christmann, P. and Taylor, G., 2001, Globalization and the Environment: Determinants of Firm Self-Regulation in China, *Journal of International Business Studies*, 32 (3): 439–458.

Corbett, C.J. and Kirsch, D.A., 2000, ISO 14000: An Agnostic's Report from the Frontline. ISO 9000 + ISO 14000 News, 9 (2): 4–17.

Corbett, C.J. and Kirsch, D.A., 2001, International Diffusion of ISO 14000 Certification, *Production and Operations Management* 10 (3): 327–342.

Dekimpe, M.G., Parker, P.M., and Sarvary, M., 2000b, Multimarket and Global Diffusion. Chapter 3 in Mahajan, V., E. Muller and Y. Wind (eds.), *New-Product Diffusion Models*, Kluwer Academic Publishers, Boston.

Delmas, M.A., 2002, The Diffusion of Environmental Management Standards in Europe and in the United States: An Institutional Perspective, *Policy Sciences*, 35: 91–119.

Delmas, M.A., 2003, In Search of ISO: An Institutional Perspective On The Adoption Of International Management Standards. Institute for Social, Behavioral, and Economic Research. ISBER Publications. Paper 2. http://repositories.cdlib.org/isber/publications/2

Guler, I., Guillén, M.F., and MacPherson, J.M., 2002, Global Competition, Institutions, and the Diffusion of Organizational Practices: The International Spread of ISO 9000 Quality Certificates, *Administrative Science Quarterly*, 47 (2): 207–232.

Hair, J.F., Tatham, R.L., Anderson, R.E., and Black, W., 1998, Multivariate Data Analysis (5th Edition), Prentice-Hall, New Jersey.

Harzing, A.-W., 1997, Response Rates in International Mail Surveys: Results of a 22 Country Study, *International Business Review*, 6 (6): 641–665.

ISO, 2004, The ISO Survey of ISO 9001:2000 and ISO 14001 Certificates – 2003 (to 31 December 2003), Available at http://www.iso.ch.

Keown, C.F., 1985, Foreign Mail Surveys: Response Rates Using Monetary Incentives, *Journal of International Business Studies*, 16 (3): 151–154.

Lee, T.Y., 1998, The Development of ISO 9000 Certification and the Future of Quality Management: A Survey of Certified Firms in Hong Kong, *International Journal of Quality & Reliability Management*, 15 (2): 162–177.

Lücke, M., 1996, Diffusion of Innovations in the World Textile Industry, *Journal of Economic Development*, 21 (2): 121–135.

Maddala, G.S., 1983, Limited-Dependent and Qualitative Variables in Econometrics, Cambridge University Press, Cambridge, UK.

Mendel, P.J., 2001, International Standardization and Global Governance: The Spread of Quality and Environmental Management Standards, In Hoffman, A. and M. Ventresca (eds.), *Organizations, Policy, and the Natural Environment: Institutional and Strategic Perspectives*, Stanford University Press.

Miller, I., Miller, M., and Freund, J.E., 1999, *John E. Freund's Mathematical Statistics*, Prentice-Hall, New Jersey.

Naveh, E., and Marcus, A.A., 2004, When does the ISO 9000 Quality Assurance Standard Lead to Performance Improvement? Assimilation and Going Beyond, *IEEE Transactions on Engineering Management*, 51 (3): 352–363.

Naveh, E., Marcus, A., and Moon, H.K., 2004, Implementing ISO 9000: Performance Improvement by First or Second Movers, *International Journal of Production Research*, 42 (9): 1843–1863.

Neumayer, E., and Perkins, R., 2004, What Explains the Uneven Take-up of ISO 14001 at the Global Level? A Panel Data Analysis, *Environment and Planning A*, 36 (5): 823–839.

Potoski, M., and Prakash, A., 2004, Regulatory convergence in nongovernmental regimes? Cross-national adoption of ISO 14001 certifications, *Journal of Politics* 66 (3): 885–905.

Scott, W.R., 1995, *Institutions and Organizations: Theories and Research*, Sage, London.

Singh, J., 1995, Measurement Issues in Cross-National Research, *Journal of International Business Studies* 26 (3): 597–619.

Terlaak, A., and King, A.A., 2005, The Effect of Certification with the ISO 9000 Quality Management Standard: A Signaling Approach, *Journal of Economic Behavior and Organization*. Forthcoming.

Terziovski, M., Samson, D., and Dow, D., 1996, The Business Value of Quality Management Systems Certification: Evidence from Australia and New Zealand, *Journal of Operations Management*, 15 August 1–18.

UNIDO, 2003, INDSTAT3 2003 ISIC Rev. 3, UNIDO, Vienna.

Risk Management in Global Supply Chain Networks

N. Viswanadham and Roshan S. Gaonkar
The Logistics Institute – Asia Pacific, National University of Singapore, 10 Kent Ridge Crescents, Singapore 119260

Abstract: In this paper, we develop a framework to classify the Global supply chain risk management problems and present an approach for the solution of these problems. The risk management problems need to be handled at three levels strategic, operational and tactical. In addition, risk within the supply chain might manifest itself in the form of deviations, disruptions and disasters. To handle unforeseen events in the supply chain there are two obvious approaches: (1) to design chains with built in risk-tolerance and (2) to contain the damage once the undesirable event has occurred. Both of these approaches require a clear understanding of undesirable events that may take place in the supply chain and also the associated consequences and impacts from these events. We focus our efforts on mapping out the propagation of events in the supply chain due to supplier non-performance, and employ our insight to develop a mathematical programming based model for strategic level deviation and disruption management. The first model, a simple integer quadratic optimization model, adapted from the Markowitz model, determines optimal partner selection with the objective of minimizing both the operational cost and the variability of total operational cost. This model offers a possible approach to robust supply chain design.

Key words: Supply Chain Risk Management; Risk Management; Supply Chain Planning; Supply Chain Design; Mean-Variance Optimization; Cause-Consequence Diagrams; Failure Analysis.

1.   Introduction

Supply chain networks are global in nature, comprising of complex interactions and flows of goods, information and funds between companies and facilities geographically distributed across countries and continents. Such chains are currently in operation in a variety of industries such as consumer electronics, automotive, pharmaceutical, aerospace, etc. Despite their complexity, most manufacturing supply chains are structurally similar. The member companies in a typical manufacturing supply chain network include the suppliers and their suppliers, assembly plants, distributors, retailers, inbound and outbound logistics providers and financing institutions. Many factors impact efficient trade in global supply chain networks: inventory visibility, late shipments, transaction costs, import and export laws compliance, customs delays, quality control problems, logistics and transport breakdowns and duplications are just a few. As China, India, Brazil, and Eastern Europe continue to grow as important economic powerhouses – with other countries becoming key second level suppliers, the cross border issues tend to dominate the supply chain management. Regulatory and compliance complexity as well as infrastructure challenges in those markets create critical supply chain and trade issues (e.g. Anti-terrorism, secure supply chain initiatives, trade facilitation etc). In fact under the intense competitive scenario prevalent today, competition is no longer between companies but between global supply chain networks with similar product offerings, serving the same global customer. The location of the supply chain constituents and the ecosystem in those countries determine the competitiveness of the supply chain. The less studied subject in the supply chain risk field is the Government risk. The antidumping duties, voluntary export restrictions are the means by which Governments would like to protect themselves against the WTO mandates and also the multilateral free trade agreements.

Because supply chain performance is inherently unpredictable and chaotic, supply chain practitioners often must seek safety mechanisms to protect against unforeseen events. Significant efforts are expended to expedite orders, to check order status at frequent intervals, to deploy inventory "just-in-case" and to add safety margins to lead times. These are some of the creative ways employed to counter the occurrence of unforeseen events. These time and material inventories along with limited communications among supply chain partners hide the problems until they lead to serious consequences. Whilst risk has always been present in the process of reconciling supply with demand, there are a number of factors, which have emerged in the last decade or so, which might be considered to have increased the level of risk. These include – a focus on efficiency rather than effectiveness; the globalization of supply chains; focused factories and centralized distribution; the trend towards outsourcing; reduction of the supplier base; volatility of demand; lack of visibility and control procedures. As a result, it has become extremely important for channel masters to employ risk management tools in the management of their supply chains.

Supply chain risk is defined by the distribution of the loss resulting from the variation in possible supply chain outcomes, their likelihood, and their subjective values. Supply chain risks comprise risks due to variations in information, material and product flows, which originate at the original supplier and lead to the delivery of the final product to the end user. Thus supply chain risks refer to the possibility and effect of a mismatch between supply and demand. Furthermore, risk consequences can also be associated with specific supply chain outcomes like supply chain costs or quality. Within this context, we can identify the following basic constructs of supply chain risk management:

1. Risk sources
2. Risk consequences
3. Risk drivers
4. Risk mitigating strategies

An increased awareness of the existence of the disturbances and their sources of origin in the supply chain may enable better preparedness for handling or preventing them.

While studying risk in a supply chain network context, one also has to remember that a supply chain comprises a network of companies that belong to an industry vertical embedded in a business and social environment. Hence, supply chains are subjected to internal risks resulting from the interaction between firms within the supply chain and to external risks that are felt by all supply chain networks in the industry, and within the same environment. Consequentially supply chain risks can arise at four levels: organizational, network level, industry level and environmental level, as elaborated in Sect. 2.1. An excellent discussion on this topic may be found in Miller (1992).

In terms of existing solutions, the existing ERP, SCM, EAI and other B2B solutions are designed to improve efficiency of the supply chains and not to enhance their reliability or robustness under uncertainty. Some vendors offer partial solutions to this problem under the name of Supply Chain Event Management (SCEM). These offerings include track and trace, supply chain visibility and alert messaging solutions (Bittner, 2000), which merely notify the human operator of unexpected occurrences and leave him to resolve the issue. In such a scenario, there is a critical need for a framework and for suitable tools that would allow companies and managers to better understand the presence and significance of various types of risks and allow them to manage it better. In this paper we attempt to address these needs from the perspective of a channel master.

1.1   Previous Work

In a very general sense, research from high reliability organizations (HROs), networked organizations, and inter-organizational systems is relevant in the study of supply chain reliability, trust and risk (Grabowski et al., 2000; Grabowski and Robots; 1999). Some of the research within this area focuses on risk management in a special breed of organizations, called virtual organizations, which are also a collection of companies under independent ownership that come together for a common purpose such as fighting forest fires or mitigating the risk of oil spills.

However, in terms of directly relevant work in the area of supply chain risk management, Paulsson (2003) provides a good survey of the recent literature in the field. Some of the commonly studied supply chain risks are disruption risk, terrorism risk and the risks from natural disasters.

With reference to disruption risks, managing such risks in global supply chains includes the following procedures: identifying sources of risk, determining the means by which such risks can take place, estimating the potential consequences, and providing the approaches to mitigating and handling these consequences. Many factors can contribute to disruption risks, including natural disasters, for example, the earthquake in Taiwan in September 21, 1999 and the SARS virus outbreak in 2003, and risks arising from purposeful organizations or individuals, such as the September 11, 2001 terrorist attack and geopolitical risks. Kleindorfer and Wassenhove (2003) have also analyzed disruption risk management in global supply chains. On supply chain security, Lee and Wolfe (2003) recently discussed the strategic approaches to improving security without jeopardizing supply chain effectiveness.

In the area of terrorism risk there has been a great deal of interest especially after the September 11, 2001 terrorist attack in the U.S. Consequential to the attacks the global business environment together with the world's political and military landscape have changed greatly and companies have reassessed common strategies for sourcing transportation, demand planning and management. Sheffi (2001) studied supply chain management under the threat of international terrorism and proposed some methods such as setting certain operational redundancies. Martha and Subbakrishna (2002) also analyzed supply chains under terrorist attacks and proposed a so-called targeting a just-in-case supply chain strategy to face the inevitable next disaster.

Another area of particular interest in supply chain risk management is that of managing risks emanating from natural disasters. Martha and Subbakrishna (2002) have investigated the impact of natural disasters on supply chains such as the earthquake in Taiwan (September 21, 1999), outbreaks of mad cow and foot and mouth diseases in Europe (Spring, 2001), and proposed the just-in-case supply chain strategy for unexpected disasters in the future. Svensson (2002) established conceptual frameworks to analyze the vulnerability in supply chains (Supply Chain Vulnerability, 2002). Svensson also provided a typology of vulnerability scenarios in supply chains based on perceived time and relationship dependencies towards both suppliers and customers (Zsidisin, 2003).

In a slightly different area one of the authors has developed a method based on process capability indices to minimize operational and performance risk through lead-time variance minimization (Garg et al., 2004). Chen and Federgruen (2000) have also, motivated by the Markowitz model studied risk management through mean-variance minimization in the context of the newsboy problem and inventory management using a base-stock policy. In addition, there are a few commercial software solutions and technology implementations to manage supply chain exceptions and events (Bittner, 2000).

Despite these publications, since the area of supply chain risk management is an emerging area of research, there are limited perspectives, theoretical models and frameworks addressing the area. We wish to provide exactly such a theoretical basis in this paper and attempt to highlight how some analytical tools can be employed to manage risk in supply chains, particularly in the context of supply risk.

1.2 Organization of this Paper

In this paper, we present a conceptual framework for the classification of supply chain risks and associated approaches to handling them. In particular, we focus on the design of robust supply chains at the strategic level through the selection of suppliers that minimize the variability of supply chain performance in terms of cost and output. In this manner we are able to build robustness into the supply chain at the planning stage itself. In Sect. 2, we present a conceptual framework for the classification of supply chain risks and associated approaches to building robustness in the supply chain. In Sect. 3, we develop models for supply chain risk management at the strategic level. In Sect. 4, we share some of our computational results and observations and finally we conclude in Sect. 5 with a discussion on the possibilities for future work.

2.   Conceptual Framework to Approach Supply Chain Risk Problems

2.1   Nature of Risk in Supply Chains

A number of business trends make supply networks more complex and global. Products and services are customised to better meet the demands of customers. Organisations have outsourced much of their activities to specialists allowing all to focus on their own core competencies. Internet based collaboration is blurring boundaries between manufacturing, logistics and distribution partners. All these trends make supply chains very efficient but also highly vulnerable to disruption. Network-related risk sources represent the second category of risk sources, which are the primary focus of this paper. These risks are of two broad kinds:

1. Firms are vulnerable not only to attacks on their own assets, but also to attacks on their suppliers, customers, transportation providers, communication lines, and other elements in their eco-system.
2. Firms are also vulnerable to irregular behavior of their network partners such as a supplier sharing sensitive product design with a competitor manufacturer.

In addition there are also risks for the industry as a whole. These risks could arise due to emergence of a disruptive technology or a new entrant with a sell direct kind of business model or due to input price, quality or quantity fluctuations. Environment related uncertainties affect businesses across all industries in a country or region. These include factors such as economic slow down, foreign exchange fluctuations, war, policy changes such as price controls, free trade zones, financial barriers, terrorist attacks and finally natural calamities such as earth quakes, storms, drought, etc.

2.2   Classification of SC Risk Problems

Based on its nature, uncertainty in the supply chain may manifest itself in three broad forms – deviation, disruption and disaster – as explained below.

2.2.1 Deviation

A deviation is said to have occurred when one or more parameters, such as cost, demand, lead-time, etc., within the supply chain system stray from their expected or mean value, without any changes to the underlying supply chain structure.

Examples of deviations:
1. Variations in demand.
2. Variations in supply.
3. Variations in procurement, production and logistics costs.
4. Variations in transportation and production lead-times.

2.2.2 Disruption

A disruption occurs when the structure of the supply chain system is radically transformed, through the non-availability of certain production, warehousing and distribution facilities or transportation options due to unexpected events caused by human or natural factors.

Examples of disruptions:
1. Disruptions in production (Taiwan earthquake resulted in disruption of IC chip production, Component production for disrupted due to a fire in Toyota's supplier's factory in Mexico resulting in downstream factory shutdown).
2. Disruptions in supply (Meat-supply was disrupted due to spread of foot-and-mouth disease in England).
3. Disruptions in logistics (US port shutdown disrupted the transportation of components from Asia to the US).

2.2.3 Disaster

A disaster is defined as a temporary irrecoverable shut-down of the supply chain network due to unforeseen catastrophic system-wide disruptions.

Examples of disasters:
1. Terrorist Action (The entire US economy was temporarily shutdown due to the downturn in consumer spending, closure of international borders and shut-down of production facilities in the aftermath of the terrorist attacks on September 11, 2001).
2. Earth quake in a supplier country such as Taiwan.

It may be noted that the classification of an event as a disruption or a disaster is dependent on the structure of a specific supply chain and its exposure to the event. Consequently, it is very likely that a particular event might manifest itself as a disruption for one supply chain network and influence another in the form of a disaster. For example, the shutdown of the US trading system, subsequent to the September 11th attacks, would be a disaster for a supply chain completely based in the US. But the same event would only be a disruption for a manufacturer, located in Asia, adopting a dual-sourcing strategy for components by procuring parts both in the US and in Europe, if the manufacturer is able to keep his supply chain running by switching from US suppliers to European ones.

In general, it is possible to design a supply chain that is robust enough to profitably continue operations in the face of expected deviations and unexpected disruptions. However, it is impossible to design a supply chain network that is robust enough to react to disasters. This arises from the constraints of any system design, which is limited by its operational specification.

Furthermore, supply chains need to be robust at three levels, strategic, tactical and operational and they need to be to handle minor regular operating deviations and major disruptions at each of these three levels. For example, at the operational level, companies require decision support systems that can act on information from various partners regarding various deviations and disruptions to reschedule activities so that the business processes are synchronized and deliveries are undertaken within customer delivery windows and cost limitations. At the tactical level, plans need to have redundancies in terms of human and machine resources and also logistics and supply organizations. At the strategic level, more reliable partners with intrinsic capabilities in deviation and disruption handling, and the skills and ability to adapt to changing market conditions will be preferred and selected. A complete classification of risk management issues, with examples, at various levels and of various scopes is presented below, with examples in Table 1.

Table 1. Types of deviations

Planning level	Type of events	Example
Strategic	Deviation	Logistics/manufacturing capacity reduction
	Disruption	Supplier bankruptcy
Tactical	Deviation	Order forecast
	Disruption	Port strike
Operational	Deviation	Lead-time variation
	Disruption	Machine/Truck breakdown

2.3 Classification of Risk Management Approaches

Accepting the fact that uncertainty cannot be completely eliminated and given that there are several possible failure modes that can affect a supply chain network; there are two choices for building "resilient supply chains": supply chains with ability to maintain, resume and restore operations after a disruption. The first approach involves the time tested "just in case" way of maintaining inventories all along the chain, employing dual or multi-sourcing and manufacturing at multiple sites. This is a highly inefficient option. A better option would be to first design a sourcing strategy taking into account the disruption costs for the most relevant failure modes and then putting in place contingency plans for each disruption that include both description of the procedures to follow and a definition of roles and responsibilities. Furthermore, within this systematic approach to risk management there can be two types of responses to manage uncertainty – preventive and interceptive.

The preventive route to managing uncertainty seeks to reduce the likelihood of occurrence of an undesirable deviation or disruption through the design of a robust chain. The process starts with identifying the set of unexpected events (also commonly known as exceptions) that can occur in the chain including the interfaces. For each of these events one can conduct the root cause analysis and devise ways and means to reduce the probability of their occurrence. One can use fault trees or fish bone diagrams for doing this. This would also enable us to compute the probability of occurrence of these undesirable exceptions.

The interceptive approach on the other hands attempts to contain the loss by active intervention subsequent to the occurrence of the event (for e.g. if there is a disruption in the supply of a critical component, buy it in an exchange). This requires a very good understanding of all the available alternatives and their impact on the supply chain.

In both cases it is first necessary to identify the exceptions that can occur in the chain, estimate the probabilities of their occurrence, map out the chain of immediate and delayed consequential events that propagate through the chain and quantify their impact. In the preventive approach, the knowledge of exception probabilities and their resulting impact is employed to design chains that are inherently robust and resilient to exceptions. In the interceptive approach, once an exception occurs, based upon the map of consequential events and their impact actions that minimize the impact of the exception are initiated (Fig. 1).

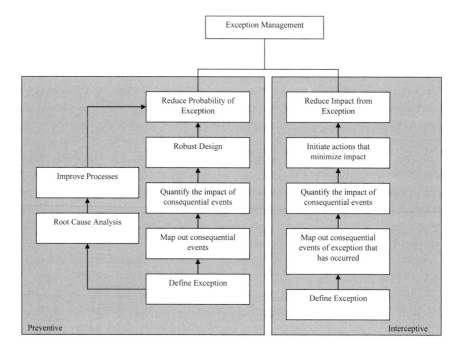

Figure 1. Exception management strategies

2.3.1 Analytical Approaches

Within the context of the broad classification of approaches suggested above a number of different analytical and computational methods and tools can be employed to design robust supply chains.

2.3.1.1 Mathematical Planning Models

Mathematical planning models can be employed to select and schedule processes and partners such that the overall supply chain is by design robust to internal and external stimuli. In particular, portfolio optimization models commonly applied in finance can be used to select a portfolio of suppliers such that the total supply chain cost variability and the consequences from supplier non-performance are within manageable limits, as demonstrated in the later sections of this paper. In addition, recent work in the area of robust optimization can also be used to generate supply chain solutions that maintain their optimality under minor deviations in environmental conditions.

2.3.1.2 Adaptive Control

A multi-level adaptive control model can be built that continuously reconfigures the supply chain such that the difference between the actual and desired performance of the supply chain is minimized. The first level of an adaptive control system can be developed from a mathematical programming-based supply chain planning model that determines optimal supply chain configurations and production and logistics schedules, which are then followed by the various participants on the supply chain. The performance of these participants is monitored and input to the second-level of the control system which then reconfigures parameters governing the first-level of the control system to provide better-designed plans that fall within the performance requirements expected from the entire supply chain. Mathematical programming models can be used to build the second-level of the control system. One such model might attempt to identify the optimal manner and location to add and deduct capacity from the supply chain such that the overall lead-times and work-in-progress inventories lie within certain specified limits. Neural networks can also be employed to build the second-level of the control system. The ensuing adaptive planning models will allow supply chains to respond in an agile manner to internal and external performance deviations.

2.4 Basics of Uncertainty Management

As mentioned in Sect. 2.3.1.2, for both preventive and interceptive approaches to risk management, it is necessary to identify the exceptions that can occur in the chain, estimate the probabilities of their occurrence, map out the chain of immediate and delayed consequential events that propagate through the chain and quantify their impact. In this context, it becomes important to identify the possible exceptions in a supply chain and their consequences before proceeding to the development of analytical models.

2.4.1 Supply Chain Exception: Definition

In attempting to analyze supply chain exceptions, our analysis here is based on a simple two-tier supply chain structure where the customer demand is directly fulfilled by a manufacturer, who in turn is supplied various components by a set of suppliers. Logistics service providers handle material movements between all the parties as shown in Fig. 2.

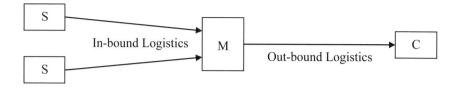

Figure 2. Simple model for analyzing exceptions

In trying to differentiate a well-executed supply chain operation from a badly managed operation we are motivated to adopt the well-accepted classical "Seven Rs" definition for the purpose of logistics, which is:

To ensure the availability of the right product, in the right quantity, in the right condition, at the right place, at the right time, at the right cost, for the right customer.

We can use this description to define a supply chain exception occurring whenever the supply chain deviates from any one of the above-required specifications either in terms of delivering the wrong product, in the wrong quantity, in the wrong condition, at the wrong place, at the wrong time, at the wrong cost and to the wrong customer. Whenever a supply chain delivery fails to stay on specification on any one of these dimensions we say that an error has been committed in that dimension.

2.4.2 Failure or Disruption Modes

In a supply chain exceptions can occur at various nodes – on the supply side, demand side, during transport or in storage – and due to a variety of different causes. There could be failures of power and communications or employee strikes. There is also a risk of breach of trust by partners, by outside elements.

In this paper, we specifically study supplier non-performance, in terms of the complete failure of a supplier to deliver components to the manufacturer or the inability of the supplier to deliver components at the promised price.

2.4.3 Cause-Consequence Diagrams

Cause-consequence diagrams or event trees are tools commonly used in reliability analysis to study the overall impact of a particular failure on the entire system. Based on the supply chain configuration, we can develop cause-consequence diagrams for each failure described above. However, given our interest in developing models for supplier selection, we employ these cause-consequence diagrams to specifically analyze the effect of supplier non-performance on the supply chain and to estimate the associated shortfalls in supply. For this purpose we develop the cause consequence diagram for supplier non-performance as given below in Fig. 3.

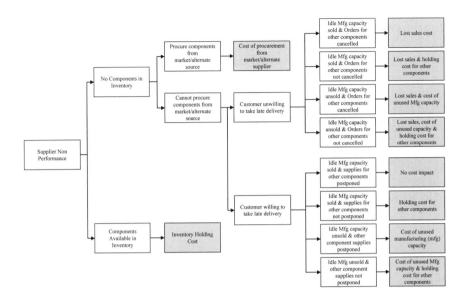

Figure 3. Cause consequence diagram for supplier non-performance and the resulting outcome

Given the probability of occurrence of the initiating event, which is supplier non-performance, and the probabilities for the various intermediary events, we can calculate the probability of occurrences for each of the end states or outcomes. Furthermore, each of these end states may result in different levels of supply shortfalls and financial cost. Hence, given the probability of each end state and the supply shortfall or financial cost for each end state, we can calculate the expected shortfall or financial risk for the non-performance of a given supplier. Such an analysis can be repeated for each supplier, and the least risky supplier can be identified as the one whose non-performance results in the least expected supply disruption or least expected financial loss.

3. Strategic Level Supply Risk Management

With the above foundation in the basics of supply chain risk management we now highlight the above approach by presenting two representative models for strategic level supply chain risk management, from the perspective of the channel master. With reference to our classification presented earlier the first model falls under the class of strategic level problems for deviation management and the second falls under the class of strategic level disruption management models. Both models employ the preventive approach to risk management based on the use of mathematical modeling techniques as described below.

1. *Strategic-level Deviation Management Model*: Given the expected costs and variability (deviation) of costs for all suppliers, the first problem relates to the selection of an optimal group of suppliers such that the expected cost of operating the entire supply chain and the risk of variations in total supply chain costs is minimized.
2. *Strategic-level Disruption Management Model*: Given the expected probabilities for various supplier disruption scenarios and the supply shortfalls under each of these scenarios the objective for the manufacturer is to choose a set of suppliers that minimize the expected shortfall during the operation of the supply chain.

It may be noted that depending on the horizon of the risk minimization and the underlying nature of the causal events the risk-parameter can be minimized using a deviation management model or a disruption management model. This choice of model will primarily depend on whether the fundamental supply chain has changed or not. If the underlying supply chain network, described by the linkages between the various supply chain participants, changes for the event studied a disruption model will be appropriate and on the other hand if the supply chain network retains its linkages a deviation model will be appropriate. This distinction between the influence of the causal events also results in certain modeling differences between the deviation management models and the disruption management models. In a deviation management model the risk-parameter to be minimized (typically a performance metric such as cost, time or demand) will invariably be modeled as a continuous variable, possibly defined by its mean and average. On the other hand, in a disruption management model the linkage impacting the supply chain network will typically be model as discrete 0 or 1 model representing the existence or absence of the linkage. It may be recognized that even while modeling disruptions the performance metrics themselves may be continuous at the system level.

In addition, for our models presented here we make the assumption that the supply chain is distributed globally and each player within the chain has its own goals, policies and cultures. The channel master who occupies a dominant position in the chain has all the information on its partners, including costs and schedules of the suppliers, the logistics providers, etc to be able to make a rational decision in the interest of minimizing risk.

3.1 Strategic Level Deviation Management Model

We propose an integer quadratic programming model for partner selection that tries to minimize the overall cost impact from the deviation in supplier costs. Such a model will be very useful to supply chain owners and channel masters. The model is an adaptation of the Markowitz model for financial portfolio management, for the purpose of managing a portfolio of suppliers. For this model, we define the impact in terms of the risk as given by the deviation of the total supply chain cost from its expected mean value. Given the expected costs and the variability of costs for all suppliers and manufacturers the objective is to choose a set of suppliers and manufacturers that minimize the expected cost of operating the entire supply chain and at the same time minimize the risk of variations in the total supply chain cost. The selection of these partners also considers the allocation of orders between these selected partners.

The mean costs and variability of the costs for each supplier can be obtained from an analysis of their historical performance or by considering the probabilities of their non-performance and the associated costs of handling the consequent impacts. Furthermore, due to the stochastic nature of events in the cause-consequence diagram we can safely assume that in general the final outcomes and associated costs of supplier non-performance will be normally distributed.

Identifiers
$m \in M$: Manufacturer identifier.
$i \in I$: Component identifier.
$s \in S_{mi}$: Supplier identifier amongst the set of suppliers for component i to a specific manufacturer m.

Parameters
C : Mean cost of the supply chain entity.
V : Cost variability for the supply chain entity.
N : Minimum number of entities to procure from.
μ :Risk aversion parameter ($0 < \mu < \infty$).

Large values for μ emphasize risk minimization and small values cost minimization.

Variables

X : Fraction of orders and hence costs allocated between manufacturers. $(0 < x < 1)$.

Y : Fraction of orders and hence costs allocated between suppliers for a specific manufacturer. $(0 < y < 1)$.

F : 0 if supply chain entity is not selected and 1 if selected.

Model

Minimize

$$\sum_{m=1}^{M} \sum_{i=1}^{I} \sum_{s=1}^{S_{mi}} y_s C_s F_s + \sum_{m=1}^{M} x_m C_m Y_m \tag{1}$$

$$+ \mu \left(\sum_{m=1}^{M} \sum_{i=1}^{I} \sum_{s=1}^{S_{mi}} y_s^2 V_s F_s + \sum_{m=1}^{M} x_m^2 V_m Y_m \right)$$

Subject to

$$\sum_{m=1}^{M} x_m F_m = 1 \tag{2}$$

$$\sum_{s=1}^{S_{mi}} y_s F_s = F_m \qquad forall \qquad m \in M \,\& \, i \in I \tag{3}$$

$$F_m \geq F_s \qquad forall \qquad m \in M \,\&\, s \in S_{mi} \tag{4}$$

$$\sum_{m=1}^{M} F_m \geq N_m \tag{5}$$

$$\sum_{s=1}^{S_{mi}} F_s \geq N_{mi} \qquad forall \qquad m \in M \,\& \, i \in I \tag{6}$$

The objective of the model is to choose manufacturers and their suppliers and allocate order quantities between them in a manner such that the expected cost of operating the supply chain is minimized and also the variability of the overall costs is minimized as well. This is subject to the constraint that the selected set of manufacturers, between them, fulfill the order (2) and that the selected set of suppliers for these manufacturers, between them, fulfill the demand for all components (3). Suppliers are part of the supply chain only when the manufacturers they supply to are involved (4). Furthermore, there might be other policies that require a minimum number of manufacturers or suppliers to be engaged at each level of the chain for the sake of redundancy and greater reliability (5) and (6).

3.2   Strategic Level Disruption Management Model

With the probabilities for supplier non-performance and knowledge of supply shortfalls under various resulting end-states (as obtained from the cause-consequence diagram), we propose a mixed integer-programming model for partner selection that tries to minimize the overall impact on the supply shortfall consequential from the exception of supplier non-performance. Such a model will be very useful to manufacturers, supply chain owners and channel masters who want to incorporate robustness into their supply chains. The model is an adaptation of the credit risk minimization model employed in financial portfolio management, for the purpose of managing a portfolio of suppliers. For this model, we define the impact in terms of the risk as given by the expected shortfall in the total supply from its expected value. Given the expected probabilities for various exception scenarios and the supply shortfalls under each of these scenarios the objective for the manufacturer is to choose a set of suppliers that minimize the expected shortfall during the operation of the supply chain.

Identifiers
$s \in S$ = Supplier identifier.
$i \in I$ = Scenario (state) identifier. I is the set of all supply scenarios (states), which is obtained as a mix of all combinations of supplier non-performance events for all the suppliers in the set J.

Parameters
K = Quantity required by the manufacturer.
x_i = Quantity supplied by supplier i.
R_j = Relation cost of including supplier j into the supply chain.
C_j = Capacity of supplier j.

Variables
F_j = 0 if supplier j is not selected and 1 if selected.
y_i = Shortfall in total supply to manufacturer in scenario i.

Model
Minimize

$$\sum_{i=1}^{I} p_i y_i + \sum_{s=1}^{S} R_s F_s \tag{1}$$

Subject to

$$K - \sum_{s=1}^{S} x_s = y_i \qquad forall \qquad i \in I \tag{2}$$

$$x_s = F_s * C_s \qquad forall \qquad s \in S \tag{3}$$

The objective of the model is to choose suppliers such that the expected shortfall in supply, in the face of supplier disruptions is minimized. This is subject to the constraint (2) which calculates the shortfall for each possible supply scenario. Also, the quantity supplied by any supplier is dependent on its capacity and also on the decision whether or not the supplier is included into the supply chain network (3). When the supplier is included into the supply chain network his supplies are equivalent to his capacity. This may be visualized as representing the capacity that is contracted or is expected to be contracted with the supplier.

4.   Computational Results

For representative purposes, both the models described above were formulated in Microsoft Excel and solved using the Solver add-in.

4.1   Strategic-Level Deviation Management Model

This model was solved for a problem with five manufacturers, dealing with five suppliers each, for each of the two components required in their manufacturing. The risk aversion factor was taken as 25 and it was required that at least two manufacturers be selected for fulfilling the orders (Table 2).

Due to the non-linear nature of the problem, the final solution obtained depends very much on the initial values of the variables. Moreover, the choice of manufacturers is the most critical decision since it also decides to a large extent the choice of suppliers. Hence, the model was solved for various initial solutions corresponding to all the possible combinations of supplier selection. The optimal solution obtained as a result is given below (Table 3).

Table 2. Cost and variance of cost for each partner

Manufacturer			Component 1			Component 2		
Sup	C	V	Sup	C	V	Sup	C	V
			S1	10	4	S1	44	7
			S2	15	3	S2	45	6
Mfg 1	90	8	S3	25	1	S3	47	5
			S4	20	2	S4	43	6
			S5	12	2	S5	45	6
			S1	13	3	S1	50	4
			S2	17	2	S2	45	6
Mfg 2	81	7	S3	19	1	S3	44	6
			S4	15	3	S4	47	5
			S5	10	3	S5	43	7
			S1	14	2	S1	42	7
			S2	16	3	S2	46	5
Mfg 3	84	8	S3	15	2	S3	49	4
			S4	11	4	S4	48	4
			S5	15	2	S5	44	6
			S1	12	3	S1	45	5
			S2	10	3	S2	45	6
Mfg 4	93	6	S3	20	3	S3	48	4
			S4	19	2	S4	46	6
			S5	18	2	S5	50	3
			S1	16	2	S1	48	5
			S2	18	2	S2	47	6
Mfg 5	99	5	S3	21	1	S3	51	4
			S4	14	2	S4	51	5
			S5	12	3	S5	48	5

C Mean cost; *V* Variance of cost; *Mfg* Manufacturer; *Sup* Supplier

Table 3. Cost and variance of cost for each partner

Manufacturers		Component 1		Component 2	
Mfg selected	Share	Sup	Share	Sup	Share
		S1	0.167	S1	0.179
		S2	0.167	S2	0.149
Mfg 4	0.46	S3	0.167	S3	0.224
		S4	0.25	S4	0.149
		S5	0.25	S5	0.299
		S1	0.176	S1	0.197
		S2	0.176	S2	0.164
Mfg 5	0.54	S3	0.353	S3	0.246
		S4	0.176	S4	0.197
		S5	0.118	S5	0.197

Sup Supplier selected,
Share = Fractional allocation of demand.

4.2   Strategic-Level Disruption Management Model

This model was solved for a problem with a single manufacturer (located in the US), dealing with five suppliers. The probabilities of supplier disruption for all the suppliers (individually and in various combination) were considered as given. The relation cost was taken as $5,000 and the quantity required by the manufacturer was 520 units. The location, capacities and risks faced for each of the suppliers is listed below in Table 4.

Table 4. Supplier pool

Supplier	Location	Capacities	Risks exposed to
Supplier 1	Ireland	250	Terrorist Attacks
			Union Strikes
Supplier 2	Taiwan	250	Earthquakes
			US East Coast Port Closure
Supplier 3	Malaysia	280	Lower Quality (Non-reliable)
			US East Coast Port Closure
Supplier 4	Singapore	340	US East Coast Port Closure
Supplier 5	USA	250	

As may be seen from Table 5, the third supplier is a non-reliable supplier based in Malaysia and the fourth a reliable supplier in Singapore, both of whom are susceptible to the risk resulting from closure of US ports. The fifth supplier is assumed to be a local supplier and is exposed to relatively insignificant risks as compared to the other four overseas-based suppliers. Based on the above characteristics of the various suppliers, the probabilities for various disruption scenarios were calculated in Table 5. Due to the lack of real-world data, our calculations are based on simulated data. However, it should be possible to perform the same analysis with detailed practical data such a country risk index and supplier rating data.

Table 5. Probabilities of various supply situations

Scenarios	Explanation	Probability
1	Supplier 1 disrupted	0.05
2	Supplier 2 disrupted	0.04
3	Supplier 3 disrupted	0.08
4	Supplier 4 disrupted	0.01
5	Supplier 5 disrupted	0.02
6	Suppliers 1 and 2 disrupted	0.0015
7	Suppliers 1 and 3 disrupted	0.0015
8	Suppliers 1 and 4 disrupted	0.0005
9	Suppliers 1 and 5 disrupted	0.0015
10	Suppliers 2 and 3 disrupted	0.0016
11	Suppliers 2 and 4 disrupted	0.0004

Continued

Scenarios	Explanation	Probability
12	Suppliers 2 and 5 disrupted	0.0008
13	Suppliers 3 and 4 disrupted	0.0008
14	Suppliers 3 and 5 disrupted	0.0048
15	Suppliers 4 and 5 disrupted	0.0001
16	Suppliers 1, 2 and 3 disrupted	0.0045
17	Suppliers 1, 2 and 4 disrupted	0.0015
18	Suppliers 1, 2 and 5 disrupted	0.0045
19	Suppliers 1, 3 and 4 disrupted	0.0015
20	Suppliers 1, 3 and 5 disrupted	0.0045
21	Suppliers 1, 4 and 5 disrupted	0.0015
22	Suppliers 2, 3 and 4 disrupted	0.0016
23	Suppliers 2, 3 and 5 disrupted	0.0032
24	Suppliers 2, 4 and 5 disrupted	0.0008
25	Suppliers 3, 4 and 5 disrupted	0.0048
26	Suppliers 1, 2, 3 and 4 disrupted	0.000045
27	Suppliers 1, 2, 3 and 5 disrupted	0.000135
28	Suppliers 1, 2, 4 and 5 disrupted	0.000045
29	Suppliers 1, 3, 4 and 5 disrupted	0.000045
30	Suppliers 2, 3, 4 and 5 disrupted	0.000032
31	All suppliers disrupted	0.00000135
32	None disrupted	0.75779665

The model was solved with the above data. The optimal selection of suppliers included Suppliers 4 and 5, with an objective value of 10,017. It might be noticed that these two suppliers are the most reliable suppliers.

5. Conclusion

We have developed a conceptual framework for the classification of risks in global supply chain networks and approaches for mitigating them. In the examples, we focus on the design of robust inbound supply chains, at the strategic level, that are resilient to deviations and disruptions that may occur at the supplier end. Our analysis is based on the identification of unforeseen events that may occur at the supplier end propagate down the supply chain leading to cost variability and supply shortfalls. Robustness is build into our supply chain design by selecting a portfolio of suppliers that minimize the variability of supply chain performance in terms of cost and output. This analysis can be extended to include other exceptions such as import and export compliance. Finally we may mention that our approach of mapping of exceptions and consequences using fault trees and event trees can form the foundation for building decision support systems for exception management in global supply chain networks.

References

M. Bittner, E-Business Requires Supply Chain Event Management, AMR Research, November 2000.

M. Grabowski, J.R.W. Merrick, J.R. Harrald, T.A. Mazzuchi and J.R. van Dorp, Risk Modelling in Distributed, Large-Scale Systems, *IEEE Transactions on Systems*, Man and Cybernetics, 30 (6), November 2000.

M. Grabowski and K.H. Robots, *Risk Mitigation in Virtual Organisations*, Organisation Science, 10 (6), Nov–Dec 1999.

U. Paulsson, Managing Risks in Supply Chains – An Article Review, *Presented at NOFOMA 2003, Oulu*, Finland, 12–13 June 2003.

Y. Sheffi, Supply Chain Management under the Threat of International Terrorism, *The International Journal of Logistics Management*, 12 (2), 1–11, 2001.

D. Garg, Y. Narahari, and N. Viswanadham, Design of Six Sigma Supply Chains, to appear in *IEEE Transactions on Automation Science and Engineering* (first issue), April 2004.

M.E. Johnson, Learning From Toys: Lessons in Managing Supply Chain Risk from the Toy Industry, *California Management Review*, 43, 106–124, 2001.

P.R. Kleindorfer and L.V. Wassenhove, *Managing Risk in Global Supply Chains*, Wharton Insurance and Risk Management Department Seminar, University of Pennsylvania, February, 2003.

H.L. Lee and M. Wolfe, Supply Chain Security without Tears, *Supply Chain Management Review*, Jan/Feb, 12–20, 2003.

Kent D. Miller, A framework for Integrated Risk Management in International Business, *Journal of International Business Studies*, second quarter, 311–331, 1992.

J. Martha and S. Subbakrishna, Targeting a Just-in-case Supply Chain for the Inevitable Next Disaster, *Supply Chain Management Review*, Sep/Oct, 18–23, 2002.

Supply Chain Vulnerability (2002). Cranfield University, UK. January, 2002.

G. Svensson, A Typology of Vulnerability Scenarios towards Suppliers and Customer in Supply Chains Based upon Perceived Time and Relationship Dependencies, *International Journal of Physical Distribution & Logistics Management*, 32, 168–187, 2002.

G.A. Zsidisin, Managerial Perceptions of Supply Risk, *Journal of Supply Chain Management*, 39, 14–25, 2003.

F. Chen and A. Federgruen, *Mean-Variance Analysis of Basic Inventory Models*, Working paper, Graduate School of Business, Columbia University, July 6, 2000.

H. Markowitz, *Portfolio Selection: Efficient Diversification of Investment*, Cowles Foundation Monograph 16, Yale University Press, New York, 1959.

Trust and Power Influences in Supply Chain Collaboration

Weiling Ke and Kwok-Kee Wei
School of Business, Clarkson University, USA
Department of Information Systems, City University of Hong Kong, Hong Kong

Abstract: There is an insufficiency of information and know-how sharing between organizations along the supply chain, though such collaboration may enhance the chain performance. In this paper, we study the factors affect the firm's predisposition to share knowledge from a socio-political perspective. Drawing on knowledge exchange and socio-political theories, we derive a model, in which trust towards the partner, in the form of competence and benevolence-based trust, and the partner's power are positively relation with the firm's predisposition to share information and know-how. In addition to their direct effects on the dependent variable, these two types of trust also moderate the relationship between various types of non-coercive power and the dependent variable. Contributions and implications of this research are discussed.

1.   Introduction

Continuous advances in information and communication technology, especially the development of Internet-based computing and communications, allow organizations to collaborate with their trading partners and integrate their supply chains (Lee and Whang, 2001). Supply chain collaboration, such as sharing information and jointly formulating demand forecast, may enable managers to stop optimizing their individual silos to work together with partners to gain global visibility across the supply chain. Thus, it makes organizations agile and responsive to market changes, ranging from customer demand to resource shortages (Lee and Whang, 2001; Sambamurthy et al., 2003). However, supply chain collaborations are not all successful. Lack of integration and information sharing are found to be the most critical factors causing the failure (Elmuti, 2002). The supply chain is still plagued with the unnecessary inventory fluctuations, often referred to as the "bullwhip effect" (Lee et al., 1997), which arises from uncoordinated channel decisions. Then what makes firms hold back in supply chain collaboration? What are the factors increasing their collaboration with trading partners? How can we help firms adopt supply chain collaboration? These are critical questions to answer before we see firms reap the great potential benefits theoretically proved by management scientists' mathematical and simulation models (e.g., Chen et al., 2000; Lee and Whang, 2001) and revealed by success stories in the industry, such as the collaboration between Wal-mart and P&G.

In this paper, we intend to seek answers to the above-mentioned questions. While supply chain collaboration may involve joint decision making and/or risk sharing, we choose to focus on the sharing of information and know-how due to two reasons. First, sharing information and know-how serves as the foundation of any type of supply chain collaboration. Second, information and know-how sharing in the industry is not enough despite that there is great potential benefits in doing so (Anand and Mendelson, 1997; Bourland et al., 1996; Chen et al., 2000; Gavirneni et al., 1999; Gerard and Marshall, 2000; Srinivasan, 2001) and the Internet has reduced its cost (Bakos, 1991; Iacovou et al., 1995; Johnston and Vitale, 1988; Majchrzak et al., 2000; Massetti and Zmud, 1996). This phenomenon seems to be inconsistent with the profit-maximization principles of organizations. To unlock the mystery, we need to look beyond the perspective of economics, such as Transaction Cost Economics (Williamson, 1985). Indeed, as argued by socio-political theory, the existence of inter-organizational relationship does not depend solely on its cost-efficiency (Hart and Estrin, 1991; Hart and Saunders, 1997; Pfeffer, 1981; Reve and Stern, 1986). Hence, we use socio-political theory as a lens to study the factors affecting the firm's predisposition to share information and know-how.

We define supply chain collaboration as practices involving sharing of know-how and information between trading partners, such as Collaborative Planning, Forecasting, and Replenishment (CPFR). Drawing upon supply chain management and social-political theories, we propose a theoretical framework on how social political factors affect the firm's predisposition to share information and know-how. We argue that trust – competence and benevolence-based – are positively related with the firm's predisposition to share information/know-how and so does non-coercive power. In addition to its direct effects, we study trust's moderating role in power and the firm's predisposition to share knowledge. This paper is organized as follow. First, we review the extant literature in supply chain collaboration and socio-political theories on trust and power. Second, we derive our conceptual model on how trust and power affect the firm's predisposition to share information and know-how. Last session is our discussion and conclusion.

2. Literature Review

2.1 Supply Chain Collaboration

Supply chain collaboration is one kind of cooperation among enterprises in the supply chain (Lee and Whang, 2001; Swaminathan and Tayur, 2003). Companies have devised many collaborative schemes to improve interaction and coordination in the chain. These include Vendor-Managed Inventory (VMI) and Co-Managed Inventory (CMI) in the retail industry, Efficient Consumer Response (ECR) in the grocery industry, Quick Response in the textile industry, Just-in-time (JIT) in manufacturing and JIT II in high technology procurement (Sheffi, 2001). Lately, the industry is moving to Collaborative Planning, Forecasting and Replenishment (CPFR), in which the trading partners jointly generate a forecast and plan for that forecast. In essence, these and many other similar initiatives are aimed at enhancing the efficiency of operation and effectiveness of decision making in the chain by ensuring trading partners share information and know-how, which are used to orchestrate the activities in supply chain, such as concerted demand forecast and replenishment.

The benefits of these supply chain collaboration models have been widely studied by management scientists (Gavirneni et al., 1999). For example, Lee and his colleagues (1997) first study the notion of lack of information sharing in the supply chain and the resulting bullwhip effect. Chen et al.

(2000) quantify the effect of forecasting and lead times on bullwhip effect under different supply chain settings. The effect of VMI systems, where buyer shares demand information with the supplier who, in turn, manages the buyer's inventory, have also been studied by researchers (e.g., Cheung and Lee, 2002). Similarly, Aviv (2001) studies the effect of sharing forecast process for production replenishment. He compares the benefits of collaborative forecasting and local forecasting in a setting with stationary demand distribution.

The great potential benefits of information and know-how sharing are attractive to companies and motivate them to enter the collaborative relationship with their trading partners. Yet supply chain collaborations are not all successful. Lack of information sharing and integration are found to be the most critical factors causing the failure (Elmuti, 2002). Thus, motivation generated by promising benefits is not sufficient for firms to actually share their information and know-how to the extent that really enhances the supply chain efficiency. Understanding what factors affect firms' predisposition to share information and know-how and how they affect such predisposition is important but lacking. The extant literature suggests that we can study the firm's supply chain collaboration from the perspective of socio-political theory, which suggests that trust and power are the major factors affecting the collaboration between organizations (Reve and Stern, 1986).

2.2   Trust

Trust, as a fundamental ingredient and lubricant and an unavoidable dimension of social interaction, has received a great deal of attention from scholars in disciplines of social psychology, sociology (Lewis and Weigert, 1985), economics (Williamson, 1991)), management (Gulati, 1995), marketing (Anderson and Weitz, 1989; Dwyer and Oh, 1987; Moorman et al., 1992), and information systems (Hart and Saunders, 1998; Jarvenpaa and Leidner, 1999; Jarvenpaa et al., 2004). Each discipline offers unique insights into the nature of trust and provides different definition of trust. Despite the differences, they all tend to agree that the concept denotes the confidence of a person, group, or organization relating to transacting with another under conditions of some uncertainty that the other's actions will be beneficial rather than detrimental to it (Kramer and Tyler, 1996; McAllister, 1995; McKnight et al., 1998). Hence, trust is an intention or willingness to depend on another party based on the optimistic anticipated behavior of another party.

The literature has come to identify multiple dimensions of trust (Mayer et al., 1995). McAllister (1995) has demonstrated empirically the importance of two types of trust: affect based and cognition based. Similarly, Mayer et al.

(1995) identify integrity, benevolence and competence as the dimensions of organizational trust. Integrity is defined as consistently adhering to a set of principles acceptable to the trustor. Benevolence, having a large affective component, is the extent to which a trustee is believed to want to do good to the trustor, aside from an egocentric profit motive. Competence refers to the skills that enable the trustee to have influence within some specific domain. Following Levin and Cross (2004), we choose to focus on competence and benevolence dimensions due to their direct relevance to the information and know-how sharing context.

Numerous scholars have acknowledged that trust can lead to cooperative behavior among organizations (Axelrod, 1984; Mayer et al., 1995; McAllister, 1995). For example, in their study of EDI adoption, Hart and Saunders (1998) found that trust is positively related with the diversity of EDI use. Similarly, goal compatibility and perception of fairness are the most robust predictors of cooperation. These relational characteristics are what trust is based on (Gulati, 1995; Ring and Van de Ven, 1994). Similarly, the results of Andaleeb's study (1995) indicate the importance of trust in explaining intentions to cooperate. In the context information sharing, trust is found to increase the amount of information exchanged (Tsai and Ghoshal, 1998). Indeed, as noted by Smith et al. (1995), "although research has identified many determinants of co-operation, virtually all scholars have agreed that one especially immediate antecedent is trust." Yet, there is not systematic research on the role of trust in supply chain management, though trade magazines have widely appeal for higher level of trust.

2.3 Power

Power is generally referred to the power holder's ability to control or influence another, the target, or to get the target to do something that s/he would not do otherwise (Emerson, 1962). In the context of inter-organizational colla-boration, Dahl defines power as the firm's capacity to influence changes of another firm, which is dependent on the firm's resources. Similarly, Weitz and Jap (1995) define power as the degree to which one party can influence another party to undertake an action that the other party would not have done.

Power is derived from the asymmetry of dependence between two parties (Pfeffer and Salancik, 1978). The greater the relative dependence, the greater the power of the less dependent firm has over the other (Blau, 1964; Emerson, 1962; Thompson, 1967). In a dyadic relationship, power is a function of (1) dependence on the other party and (2) the use of dependence to leverage change in accord with the intentions of the less dependent firm. Thus, the source and extent of relative dependence are determinants of power, which represents the firm's capacity to influence changes in another firm (Dahl, 1957; Emerson, 1962).

Hunt and Nevin (1974) dichotomized exercised power into coercive and non-coercive influencing strategies. Coercive power is the mechanism for gaining target compliance that references or mediates negative consequences of noncompliance (Roering, 1977). By contrast, non-coercive power is the mechanism that references or mediates positive consequences for compliance (Hunt and Nevin, 1974). Researchers have applied the power literature to the analysis of marketing channel relationships and found that the different types of power affect inter-organizational relationship in significant, yet contrasting ways – coercive power lowers genuine commitment by the target while non-coercive power increases commitment and the target's willingness to cooperate (Brown et al., 1995; Hunt et al., 1987).

In recent years, the firm has placed greater emphasis on developing long-term relationships to obtain substantial benefits. It chooses to use non-coercive influence strategies rather than coercive approaches due to the coercive power's increasing conflicts and negative impact on the partners' satisfaction (Frazier and Rody, 1991). Especially, when it comes to exchanging knowledge electronically through IOS, due to the difficulties in monitoring and evaluating the other party's sharing behavior, coercive power may cause great damage to the firm (Kumar and Dissel, 1996). Indeed, Hart and Saunders (1998) found that coercive power did not lead the firms to share more information through EDI. Thus, research on inter-organizational collaboration has focused on the effect of non-coercive power lately.

The existing literature suggests that there are five different types of non-coercive power – information, reward, expert, referent and legitimate (Brown et al., 1995). Information power refers to one firm's influencing its partner by providing information that can facilitate its complying. Reward power means promising to provide reward to the target if it complies. Expert power refers one firm's giving its partner the perception that it holds expertise or information that is valued by the partner, while referent power implies making the partner desire for identification with another for recognition. Legitimate power refers that the target believes in the right of partner to wield influence.

3.   Theoretical Model

As sharing of information and know-how, such as forecasting technique, is expected to affect the supply chain performance differently, it is important to pay attention to the complexity involved in these two types of sharing.

Existing research suggests a distinction between relatively simple information sharing and know-how exchange. Conceptually, these two forms of sharing involve different types of knowledge. Information is the "knowledge which can be transmitted without loss of integrity once the syntactical rules required for deciphering it are known" (Kogut and Zander, 1992). By contrast, know-how is "the accumulated practical skill or expertise that allows one to do something smoothly and efficiently" (Von Hippel, 1988). In this paper, we consider the difference between information and know-how.

In addition, we conduct this research from the more dependent firm's perspective and study how trust and power affect this firm's predisposition to share information and know-how. In particular, we study the two dimensions of trust – competence and benevolence-based trust, instead of treating trust as a first-order construct. Also, we study the non-coercive power's effect on the firm's predisposition to share knowledge and the moderating role of knowledge's tacitness in competence-based trust and knowledge sharing predisposition.

3.1 Trust

While collaborating with trading partners may enhance the chain's performance, sharing information and know-how is not without risk. Sharing proprietary information and know-how with the trading partner can be a double-edged sword, especially when such sharing is through inter-organizational system. Electronic access to the information of trading partner makes organizational boundaries more permeable. It allows the internal activities of the firm to become more transparent to its partners and the partners may use the more readily accessible knowledge in ways that the firm is not able to control (Hart and Saunders, 1997). In addition, this sharing has certain hybrid characteristics arising from the paradox of combining cooperation and competition. Though participants may attempt to formulate common goals, the objectives of collaboration cannot be wholly complementary. Thus, sharing information and know-how with trading partners introduces new sources of vulnerability.

When there is any risk involved in inter-organizational linkage setting up, trust, as the most important factor affecting cooperation, is needed to overcome the psychological barrier imposed by risk. We argue that trust directly affects the firm's predisposition to share information and know-how. Benevolence-based trust's effect can be in three aspects. First, trusting the partner to be benevolent should alleviate the firm's anxiety that the partner may abuse and/or disclose information and know-how to a third party. It reduces the chance of setting up control and/or monitoring system by the firm and thereby decreases the cost of information and know-how sharing (Zaheer et al., 1998). Secondly, this trust makes the firm believe that the partner will share accurate and useful information and know-how, which is beneficial to supply chain performance (Levin and Cross, 2004). Such belief motivates the firm to share its proprietary information and know-how. Thirdly, trusting that the partner cares about its interests should make the firm expect fair appropriation of gains derived from supply chain collaboration (Larsson et al., 1998). Such expectation makes the firm willing to share information and know-how. Hence we have the following proposition:

Proposition 1 Benevolence-based trust is positively related with the firm's predisposition to share information and know-how with its partner.

Trusting in the partner's competence should also affect the firm's predisposition to share information and know-how. The partner's competence makes the firm perceive high usefulness of information and know-how shared by the partner. This perception leads the firm to be willing to exchange its proprietary knowledge for the partner's expertise (Dirks and Ferrin, 2001; Kramer, 1999; Tyler et al., 1996). Also, competence-based trust makes the firm think positively about the partner's business sense and judgment, and expect optimistically about benefits gained from collaborating with this partner. Hence, we have the following proposition:

Proposition 2 Competence-based trust is positively related with the firm's predisposition to share information and know-how with its partner.

The impact of competence-based trust on the firm's predisposition to share knowledge may also be contingent on the type of knowledge shared. The distinction between information and know-how sharing has organizational implications. The coordination required for information sharing is typically simple. By data transmission over the value-added network or Internet, for example, is straightforward. As information tends to be explicit or at least codifiable, its sharing is a matter of verbal or written communication (Kogut and Zander, 1992). By contrast, know-how involves a greater scope of activities and higher-level organizational principles. It requires extensive and dedicated coordination, as personnel interact both within and across firms for sustained periods of time. This renders know-how sharing particularly costly (Carley, 1991; Simonin, 1999; Szulanski, 1996). Thus it requires the firm to integrate the know-how sharing relationship into its own strategic framework and competence-based trust plays a central role in this motion (Jonson et al., 1998; Madhok, 1995). Trusting the partner's competence makes the firm think it can rely on the partner and learn from the partner. So the firm will be more inclined to think the relationship as a strategic asset and consider this relationship in its strategic planning. Hence we posit that when the knowledge shared is stickier – in the case of know-how – competence-based trust has a greater impact on the firm's predisposition to share knowledge.

Proposition 3 Competence-based trust is more important to the firm's predisposition to share know-how.

3.2 Power

Power refers to the less dependent firm's ability to influence the more dependent party's decision making. Though there are multiple conceptualizations of power in the extant literature, we focus on the power derived from the relative dependence in a dyadic relationship. Mathematically, the source of power can be calculated as the difference between the dependence of firm A on B and the dependence of firm B on A. If the later is higher then firm A has power over firm B (Pfeffer, 1992). When the two firms are symmetrically dependent on each other, neither party has power over the other.

Following the recent literature, we focus on the effects of non-coercive power – expert, information, referent, reward and legitimate power. Expert power makes the firm perceive that the partner holds expertise (Hunt et al., 1987), such as demand forecasting skills. Sharing information and know-how with this partner can be taken as an opportunity to benefit from this partner's expertise. Thus, it should increase the firm's predisposition to share knowledge. Similarly, information power is positively related with such predisposition. The partner can provide two types of information – information on the benefits of sharing knowledge and how to collaborate. Providing the firm with the information on benefits of knowledge sharing makes the firm realize that both firms can achieve synergistic effects of combining information and know-how, while the information on how to collaborate helps the firm tackle the technical and management problems. Both types of information facilitate the firm to embrace sharing information and know-how with the partner.

In addition, the partner may exercise its referent power and make the firm comply with the idea of sharing information and know-how (Frazier and Summers, 1986; Gaski and Nevin, 1985). Referent power influences the firm by its desire for identification with the partner. Collaborating with the partner is recognized as one kind of association with the partner. So it increases the firm's predisposition to share information and know-how. The partner can also influence the firm by its reward power. Promising economically reward the firm, the partner gives incentives for the firm to comply. The reward promise helps the firm reduce its anxiety caused by the uncertainties of gaining from the information and know-how sharing. In addition, the partner's legitimate power is another power base that can be used to affect the firm's predisposition to share information and know-how. The legitimate power makes the firm believe in the right of the partner to wield influence and comply with the partner's proposal for sharing information and know-how. Hence we expect that the partner's referent, reward and legitimate power have positive effects on the firm's predisposition to share knowledge.

Proposition 4 The partner's power is positively related with the firm's predisposition to share information and know-how.

3.3 Interaction Between Trust and Power

We have posited that trust and power are positively related to the firm's predisposition to share information and know-how. As trust reduces the complexity and uncertainty associated with the powerful partner's actions, it makes the firm confident about the collaboration and inclined to be influenced by the partner (Andaleeb, 1995; Dwyer et al., 1987). Hence, trust serves to facilitate the effects of power on the firm's predisposition to share information and know-how. The moderating role of trust is not new, but it has received scant attention from researchers in inter-organizational relationships.

Competence-based trust towards the partner should increase the chance that the target will be willing to accept the influence of information, expert, referent and legitimate power of the partner and thereby be more inclined to share knowledge. When the powerful partner presents the potential benefits of supply chain collaboration, with the confidence about the partner's competence, the firm perceives high level of usefulness, accuracy and reliability of this information. It will be more willing to agree with the partner on the optimistic prospect of collaboration. Such optimistic expectation leads the firm to share information and know-how with this partner. By contrast, with low level of competence-based trust, the firm questions about accuracy and reliability of the information and know-how provided by the partner. It may not see eye to eye with the partner on the possible gains that can be derived from sharing knowledge. Thus, it is less inclined to share proprietary information and know-how with the partner.

Similarly, high level of competence-based trust facilitates the effects of expert power on chain collaboration. Trusting the partner's competence makes the firm expect that the powerful partner has the expertise on how to reap the benefits promised by supply chain collaboration. The perceived uncertainties associating with collaboration with this capable partner is decreased. Hence, with high competence-based trust, the same level of expert power will have a larger effect on the firm's predisposition to share information and know-how.

Likewise, with higher confidence on the partner's competence, the firm expects higher chance of success in collaborating with this partner and thus more willing to comply with the partner's request for information and know-how sharing. Hence competence-based trust also facilitates referent and legitimate power's effect on firm's willingness to collaborate (Table 1). We have the following proposition:

Proposition 5 Competence-based trust moderates the relationship between the partner's information, expert, referent and legitimate power and the firm's predisposition to share information and know-how.

A state of asymmetric dependence makes the firm endure the fear of being exploited (Geyskens et al., 1996). Such negative feelings can be mitigated by the firm's benevolence-based trust towards the partner. In addition to reducing the firm's perceived risk associated with its partner's actions, benevolence-based trust leads the firm to attribute cooperative and sincere intentions to its partner (Andaleeb, 1995). Hence, benevolence-based trust may increase the chance that the firm will be willing to be influenced by the partner and collaborate with this partner.

The firm's benevolence-based trust towards the powerful partner facilitates the information power's effect on the firm's predisposition to share information and know-how. When the partner is providing information regarding the benefits of supply chain collaboration, the firm is vulnerable to the benevolence of the partner (Lee, 1997). With high level of benevolence-based trust, the firm is more likely to perceive that the partner is promoting joint interests constructively by providing this information (Dwyer et al., 1987). Thus, it takes more "face value" of this information and is more willing to be influenced by this partner.

In addition, the firm's benevolence-based trust moderates the relationship between the partner's reward power and the firm's predisposition to share information and know-how. With high level of benevolence-based trust, the firm perceives a positive orientation of the reward power influence exercised by the powerful partner. Also, the trust towards this partner makes the firm believe that the partner will keep its promise and be fair with the benefit appropriation. Hence, with higher level of benevolence-based trust, the partner's reward power has a greater effect on the firm's supply chain predisposition to share knowledge (Table 1).

Proposition 6 Benevolence-based trust facilitates the effect of partner's information and reward power on the firm's predisposition to share information and know-how.

Table 1. Moderating role of trust between power and predisposition to share knowledge

	Information	Expert	Reward	Referent	Legitimate
Competence	+	+		+	+
Benevolence	+		+		

4.   Discussion and Conclusion

From a social-political perspective, this paper studies the factors affecting the firm's predisposition to sharing information and know-how with its trading partner. Specially, we study the effects of two types of trust – competence and benevolence. While recognizing that both of these two types of trust are positively related with the firm's predisposition to collaborate with its partner, we argue that competence-based trust is more important for know-how sharing. In addition, we contend that the five types of non-coercive power have direct impact on the firm's predisposition to share knowledge. Moreover, we study the moderating role of trust in the relationship between different types of power and the firm's predisposition to share information and/or know-how.

Limitations of this study should be acknowledged before we discuss its contribution and implication. First, we conduct this study from the perspective of the more dependent firm. Dyadic study may shed more new lights on the issue under study. Especially, a longitudinal study on the interaction between dyadic firms in sharing information and know-how may provide significant insights. Secondly, we focus on the effects of trust and power, though there are other social-political factors that may affect the firm's predisposition, such as complementary goals and shared values. Our suspect is these factors may serve as antecedents of trust and play moderating role between power and the firm's predisposition to share knowledge. Further research should be conducted to extend our conceptual model. Thirdly, empirical studies should be conducted to test the validity of the conceptual model proposed in this paper.

Within these limitations in mind, this study's theoretical contribution is to both the supply chain management and knowledge exchange literatures. To the supply chain management, we propose a conceptual model that integrates the understanding of how trust and power affect the firm's predisposition to share information and know-how. While trust is recognized as a critical factor affecting the success of supply chain collaboration, there is little systematic study on its impact in this area. Similarly, power remains a predominant yet overlooked factor in supply chain strategy. Thus, our study addresses the important and understudied issues in supply chain management. While drawing on theories in inter-organizational collaboration, we extend the existing models by studying the moderating role of different types of trust in the relationship between power and the firm's sharing predisposition. We also contribute to the knowledge exchange research area. Most of extant knowledge exchange works are on the interaction between individuals within the organization (Brandon and Hollingshead, 2004; Jarvenpaa et al., 2004; Levin and Cross, 2004). As deriving competitive advantages by cooperating with other organizations is important, research on inter-organizational knowledge exchange is imperative, given knowledge is the critical resource in this era. Also, different from the other studies on inter-organizational knowledge exchange (Helper, 1990; Uzzi, 1999; Zaheer and Venkatraman, 1995), we incorporate the effect of power, which is a critical factor affecting organizations' behavior and decision making.

Finally, we feel our work holds significance for practitioners. With the popularization of the concept of relational-based competitive advantage, there has been an increased interest among practitioners in the role of trust and power in inter-organizational relationships, the supply chain collaboration in particular. Our research offers two main insights that can be helpful to practitioners. First, most firms are not ware of the broad scope of power dimensions and therefore may not actively manage their own power bases. Awareness of this can help executives exercise appropriate influencing strategies and get its partners comply with supply chain collaboration. Second, our study argues that trust moderates power's effect on the firm's predisposition to share information and know-how, in addition to its direct effect on the dependent variable. Our viewpoint sheds light on the importance of trustworthiness for supply chain collaboration. Practitioners may find it fruitful to focus on ways to improve trust as a pragmatic way to improve the needed sharing of information and know-how.

References

Anand, K.S. and Mendelson, H., 1997, Information and organization for horizontal multimarket coordination, *Management Science*, 43 (12): 1609–1627.

Andaleeb, S.S., 1995, Dependence relations and the moderating role of trust: implications for behavioral intentions in marketing channels, *International Journal of Research in Marketing*, 12 (2): 157.

Anderson, E. and Weitz, B., 1989, Determinants of Continuity in Conventional Industrial Channel Dyads, *Marketing Science*, 8 (4): 310.

Aviv, Y., 2001, The effect of collaborative forecasting on supply chain performance, *Management Science*, 47 (10): 1326–1343.

Axelrod, R., 1984, *The Evolution of Cooperation*, Basic Books, New York.

Bakos, J.Y., 1991, Information links and electronic marketplaces: the role of interorganizational information systems in vertical markets, *Journal of Management Information Systems* 8 (2): 31–52.

Blau, P.M., 1964, *Exchange and power in social life*, Wiley, New York.

Bourland, K.E., Powell, S.G., and Pyke, D.F., 1996, Exploiting timely demand information to reduce inventories, *European Journal of Operational Research*, 92 (2): 239.

Brandon, D.P. and Hollingshead, A.B., 2004, Transactive memory systems in organizations: matching tasks, expertise, and people, *Organization Science*, 15(6): 633–644.

Brown, J.R., Johnson, J.L., and Koenig, H.F., 1995, Measuring the sources of marketing channel power: A comparison of alternative approaches, *International Journal of Research in Marketing*, 12 (4): 333.

Carley, K., 1991, A theory of group stability, *American Sociological Review*, 56 (3): 331–354.

Chen, F., Drezner, Z., Ryan, J.K., and Simchi-Levi, D., 2000, Quantifying the bullwhip effect in a simple supply chain: the impact of forecasting, lead times, and information, *Management Science*, 46 (3): 436–443.

Cheung, K.L. and Lee, H.L., 2002, The inventory benefit of shipment coordination and stock rebalancing in a supply chain, *Management Science*, 48 (2): 300–306.

Dahl, R.A., 1957, The concept of power, *Behavioral Science*, 2: 201–218.

Dirks, K.T. and Ferrin, D.L., 2001, The role of trust in organizational settings, *Organization Science*, 12 (4): 450–467.

Dwyer, F.R. and Oh, S., 1987, Output sector munificence effects on the internal political economy of marketing channels, *Journal of Marketing Research*, 24 (4): 347–348.

Dwyer, F.R., Schurr, P.H., and Oh, S., 1987, Developing buyer-seller relationships, *Journal of Marketing*, 51 (2): 11.

Elmuti, D., 2002, The perceived impact of supply chain management on organizational effectiveness, *Journal of Supply Chain Management*, 38 (3): 49–57.

Emerson, R.M., 1962, Power-dependence relations, *American Sociological Review*, 27 (1): 31–41.

Frazier, G.L. and Rody, R.C., 1991, The use of influence strategies in interfirm relationships in industrial product channels, *Journal of Marketing*, 55 (1): 52.

Frazier, G.L. and Summers, J.O., 1986, Perceptions of interfirm power and its use within a franchise channel of distribution, *JMR, Journal of Marketing Research*, 23 (2): 169.

Gaski, J.F. and Nevin, J.R., 1985, The differential effects of exercised and unexercised power sources in a marketing channel, *JMR, Journal of Marketing Research*, 22 (2): 130.

Gavirneni, S., Kapuscinski, R., and Tayur, S., 1999, Value of information of capacitated supply chains, *Management Science*, 45(1): 16–24.

Gerard, P.C. and Marshall, F., 2000, Supply chain inventory management and the value of shared information, *Management Science*, 46 (8): 1032–1048.

Geyskens, I., Steenkamp, J.-B.E.M., Scheer, L.K., and Kumar, N., 1996, The effects of trust and interdependence on relationship commitment: a trans-Atlantic study, *International Journal of Research in Marketing*, 13 (4): 303–317.

Gulati, R., 1995, Does familiarity breed trust? The implications of repeated t, *Academy of Management Journal*, 38 (1): 85–112.

Hart, P., and Estrin, D., 1991, Inter-organization networks, computer integration, and shifts in interdependence: the case of the semiconductor industry, *ACM Transactions on Information Systems*, 9 (4): 370–398.

Hart, P., and Saunders, C., 1997, Power and trust: critical factors in the adoption and use of electronic data interchange, *Organization Science,* 8 (1): 23–42.

Hart, P.J., and Saunders, C.S., 1998, Emerging electronic partnerships: antecedents and dimensions of EDI use from the supplier's perspective, *Journal of Management Information Systems*, 14 (4): 87–111.

Helper, S., 1990, Comparative supplier relations in the U.S. and Japanese auto industries: an exit voice approach, *Business Economic History*, 19: 153–162.

Hunt, K.A., Mentzer, J.T., and Danes, J.E., 1987, The effect of power sources on compliance in a channel of distribution: a causal model, *Journal of Business Research*, 15 (5):377.

Hunt, S.D., and Nevin, J.R., 1974, Power in a channel Of distribution: sources and consequences, *Journal of Marketing Research*, 11: 186–193.

Iacovou, C.L., Benbasat, I., and Dexter, A.S., 1995, Electronic data interchange and small organizations: adoption and impact of technology, *MIS Quarterly*, 19 (4): 465.

Jarvenpaa, S.L., and Leidner, D.E., 1999, Communication and trust in global virtual teams, *Organization Science*, 19 (6): 791–815.

Jarvenpaa, S.L., Shaw, T.R., and Staples, D.S., 2004, Toward contextualized theories of trust: the role of trust in global virtual teams, *Information Systems Research*, 15 (3): 250–267.

Johnston, H.R., and Vitale, M.R., 1988, Creating competitive advantage with interorganizational Inf, *MIS Quarterly*, 12 (2): 153.

Jonson, J.L., Sakano, T., Voss, K., and Takenouchi, H., *Marketing Performance in U.S.-Japanese Cooperative Alliances: Effects of Multiple Dimensions of Trust and Commitment in the Cultural Interface*, Washington State University, Pullman), 1998.

Kogut, B., and Zander, U., 1992, Knowledge of the firm, combinative capabilities, and the replication of technology, *Organization Science*, 3 (3): 383–397.

Kramer, R.M., 1999, Trust and distrust in organizations: emerging perspectives, enduring questions, *Annual Review of Psychology*), 50: 569–598.

Kramer, R.M., and Tyler, T.R., 1996, *Trust in Organizations: Frontier of Theory and Research*, Sage, Thousand Oaks, CA.

Kumar, K., and Dissel, 1996, H.G.V. Sustainable collaboration: managing conflict and cooperation in interorganizational systems, *MIS Quarterly*, 20 (3): 279–300.

Larsson, R., Bengtsson, L., Henriksson, K., and Sparks, J., 1998, The interorganizational learning dilemma: collective knowledge development in strategic alliances, *Organization Science*, 9 (3): 285–305.

Lee, F., 1997, When the going gets tough, do the tough ask for help? Help seeking and power motivation in organizations, *Organizational Behavior and Human Decision Processes*, 72: 336–363.

Lee, H.L., Padmanabhan, P., and Whang, S., 1997, Paralyzing curse of the bullwhip effect in a supply chain, *Sloan Management Review*, 93–102.

Lee, H.L., and Whang, S. E-business and supply chain integration, 2001. In "The Practice of Supply Chain Management: Where Theory and Application Converge", edited by Harrison, T., Lee, H., and Neale, J., Springer New York, 2004.

Levin, D.Z., and Cross, R., 2004, The strength of weak ties you can trust: the mediating role of trust in effective knowledge transfer, *Management Science*, 50 (11): 1477–1490.

Lewis, J.D., and Weigert, A., 1985, Trust as a social reality, *Social Forces*, 63: 967–985.

Madhok, A., 1995, Revisiting multinational firms tolerance for joint ventures – a trust-based approach, *Journal of International Business Studies*, 26 (1): 117–137.

Majchrzak, A., Rice, R.E., Malhotra, A., King, N., and Ba, S., 2000, Technology adaptation: The case of a computer-supported inter-organizational virtual team, *MIS Quarterly*, 24 (4):569–600.

Massetti, B., and Zmud, R.W., 1996, Measuring the extent of EDI usage in complex organizations: strategies and illustrative examples, *MIS Quarterly*, 20 (3): 331–345.

Mayer, R.-C., Davis, J.-H., and Schoorman, F.D., 1995, An integrative model of organizational trust, *Academy of Management Review*, 20 (3): 709–734.

McAllister, D.J., 1995, Affect- and cognition-based trust as foundations for interpe, *Academy of Management Journal*, 38 (1): 24–59.

Mcknight, D.H., Cummings, L.L., and Chervany, N.L., 1998, Initial trust formation in new organizational relationships, *Academy of Management Review*, 23 (3): 473–490.

Moorman, C., Zaltman, G., and Deshpande, R., 1992, Relationships between providers and users of market research, *JMR, Journal of Marketing Research*, 29 (3): 314.

Pfeffer, J. *Power in Organizations*, Pitman Publishing Company, Marshfield, MA, 1981.

Pfeffer, J., 1992, Understanding Power in Organizations, *California Management Review*, 34 (2): 29.

Pfeffer, J., and Salancik, G.R. *The External Control of Organizations: A Resource Dependence Perspective*, Harper & Row, New York, 1978.

Reve, T., and Stern, L.W. The relationship between interorganizational form, transaction climate, and economic performance in vertical interfirm dyads, In *Marketing Channels*, L. Pellegrini and S.K. Reddy (eds.), Lexington Books, Lexington, MA, 1986, pp. 75–102.

Ring, P.S. and Van de Ven, A.H., 1994, Developmental processes of cooperative inter-organizational relationships, *The Academy of Management Review*, 19 (1): 90–118.

Roering, K.J., 1977, Bargaining in distribution channels, *Journal of Business Research*, 5 (1): 15–26.

Sambamurthy, V., Bharadwaj, A., and Grover, V., 2003, Shaping agility through digital options: reconceptualizing the role of information technology in contemporary Firms1, *MIS Quarterly*, 27 (2): 237.

Sheffi, Y., 2001, Supply chain management under the threat of international terrorism, *Journal of Logistics Management*, 12 (2): 1–11

Simonin, B.L., 1999, Ambiguity and the process of knowledge transfer in strategic alliances, *Strategic Management Journal*, 20: 595.

Smith, K.G., Carroll, S.J., and Ashford, S.J., 1995, Intraorganizational and interorganizational cooperation: toward a research agenda, *Academy of Management Journal*, 38, 7–23.

Srinivasan, R., 2001, Information sharing in a supply chain: a note on its value when demand is nonstationary, *Management Science*, 47 (4): 605–610.

Swaminathan, J.M., and Tayur, S.R., 2003, Models for supply chains in e-business, *Management Science*, 49 (10): 1387–1406.

Szulanski, G., 1996, Exploring internal stickiness: impediments to the transfer of best practice within the firm, *Strategic Management Journal*, 17(Winter): 27.

Thompson, J.D., 1967, *Organizations in Action; Social Science Bases of Administrative Theory*, McGraw-Hill, New York.

Tsai, W., and Ghoshal, S., 1998, Social capital and value creation: The role of intrafirm networks, *Academy of Management Journal*, Aug: 464–476.

Tyler, T., Degoey, P., and Smith, H., 1996, Understanding why the justice of group procedures matters: A test of the psychological dynamics of the group-value model, *Journal of Personality and Social Psychology*, 70 (5): 913.

Uzzi, B., 1999, Embeddedness in the making of financial capital: How social relations and networks benefit firms seeking financing, *American Sociological Review*, Aug: 481–505.

Von Hippel, E., 1988, *The sources of Innovation*, Oxford University Press, New York.

Weitz, B.A., and Jap, S.D., 1995, Relationship marketing and distribution channels, *Journal of the Academy of Marketing Science*, 23 (4): 305–320.

Williamson, O.E., 1985, *The economic Institutions of Capitalism*, Free Press, New York.

Williamson, O.E., 1991, Comparative economic organization: the analysis of discrete, *Administrative Science Quarterly*, Jun: 269–296.

Zaheer, A., McEvily, B., and Perrone, V., 1998, Does trust matter? Exploring the effects of interorganizational and interpersonal trust on performance, *Organization Science*, 9 (2): 141–159.

Zaheer, A., and Venkatraman, N., 1995, Relational governance as an interorganizational strategy: An empirical test of the role of trust in economic exchange, *Strategic Management Journal*, 16 (5): 373.

Foreign Direct Investment or Outsourcing: A Tax Integrated Supply Chain Decision Model

N. Viswanadham and Kannan Balaji
Indian School of Business, Hyderabad 500032, India
The Logistics Institute Asia Pacific, The National University of Singapore, Singapore-119260

Abstract: Multinational companies should look at tax information at a strategic level. Current literature on supply chain optimization does not emphasize on tax, unfortunately, to make more realistic decisions about where to make, source, locate, move and store products. Many developing economies, specifically, Asian countries, have included tax-holidays in their export-import (EXIM) policy for companies operating in Free Trade Zones (FTZs). Including this information in the supply chain planning could save millions of dollars for the companies operating globally. In this paper, we propose a tax integrated mixed integer model, for optimally deciding the FDI-outsourcing alternatives at the various stages of a global supply chain. We empirically analyze the proposed model by incorporating FTZs on an 8-stage supply chain.

1.   Introduction

Trade liberalization and information technology development accelerates firms to trade and invest across national borders. Firms could trade across national borders either by intra-firm-trade (FDI) or arms-length trade (foreign outsourcing). FDI includes corporate activities such as building plants or subsidiaries in foreign countries, and buying controlling stakes or shares in foreign companies. It is now a competitive requirement that businesses invest all over the globe to access markets, technology, and talent. Firms located in industrialised countries pursue vertical disintegration of their production processes by outsourcing some stages in foreign countries where economic conditions are more advantageous. A firm that chooses to keep the production of an intermediate input within its boundaries can produce it at home (standard vertical integration) or in a foreign country (FDI). Alternatively, a firm may choose to outsource the production to a supplier in the home country (domestic outsourcing) or in a foreign country (foreign outsourcing). Intel Corporation provides an example of the FDI strategy; it assembles most of its microchips in wholly-owned subsidiaries in China, Costa Rica, Malaysia, and the Philippines. On the other hand, Nike provides an example of foreign outsourcing strategy; it subcontracts most of its manufacturing to independent producers in Thailand, Indonesia, Cambodia, and Vietnam.

FDI and outsourcing had been studied extensively in the economics literature. Economists had developed theoretical models for investigating the decision of the firms to source abroad either through foreign outsourcing (FO) or foreign direct investment (FDI) (Antras and Helpman (2004)). In Grossman and Helpman (2002, 2005), the authors had studied the trade-off between outsourcing and in-house production in a closed economy. Instead, in Grossman and Helpman (2003) they had studied the trade-off between FDI and outsourcing in a foreign country. They assume that the producers of final goods, located in a Northern region, find it convenient to buy inputs from a Southern region, since wages in the South are lower than wages in the North. In addition, Grossman and Helpman (2003) suppose the local suppliers in South to be more efficient with respect to a production unit eventually set up in the Southern region by the final producers through a vertical FDI. However, the eventual relationship with the suppliers is plagued with contractual difficulties, linked to the uncertain legal framework of the South, and therefore for the final producers a trade-off arises between the greater efficiency gained through outsourcing, and the contract incompleteness they might avoid if they *FDI or Outsourcing: A Tax Integrated Decision Model* 3 produce their required inputs through a FDI. The work by Altomonte and Bonassi (2004) contributes with some refinements to the Grossman and

Helpman (2003) model as far as the treatment of the FDI alternative is concerned and explores the extent to which the production strategies of the final producers are sensitive to the degree of contract incompleteness of a host country, and how in turn the latter affects the establishment of linkages between the final producers and the local suppliers. Gorg et al. (2004) had done an econometric study on outsourcing using Irish manufacturing plant data. For more details on FDI and outsourcing studies we refer to Antras (2003), Domberger (1998), Feenstra (1998), Feenstra and Hanson (2001), Groot (2001), Helpman et al. (2004) and Hummels et al. (2001).

1.1   Contribution

Globalization, cost pressures and market demands for new and innovative products are key factors behind many complex supply chain challenges today. When planning a global supply chain, understanding and effectively managing tax liabilities can result in tens or hundreds of millions of dollars in savings (Irving et al. (2005)).

Standalone supply chain initiatives, such as network optimization, strategic sourcing, and lean manufacturing, reduce operating expenses and working capital requirements, as well as improve cash flow and asset utilization. They can also lead to the development of new intangible assets and improved profits. Yet because standalone supply chain initiatives focus only on pre-tax cost reduction, they overlook the fact that for each dollar of operating savings generated, only a limited portion of the benefit as little as sixty cents on the dollar, depending on the tax jurisdiction will be the actual reduction in cost after taxes.

Similarly, when tax planning is performed independently from supply chain planning, it may lead to suboptimal strategies with respect to operating cost and profit.

Either type of initiative, undertaken in isolation, prevents companies from achieving a greater after-tax return from their supply chain improvements. Conversely, when the two initiatives are integrated the combination can achieve better results. Companies can enjoy the expanded benefits of enhanced supply chain profitability and lower compliance risks without the burden of high tax rates or exposure to tax compliance risks.

In this research we propose a mixed integer model for deciding the optimal FDI-outsourcing alternatives at the various stages of a global acyclic supply chain by taking into account the export and import tax four liabilities. This model is termed as the tax integrated model. The tax integrated model is obtained as an extension of the weighted version of the model proposed in Viswanadham and Balaji (2005a). So, it is a quantitative model. That is, it would output what percentage to make or source using a particular FDI-outsourcing alternative. Integration of taxes and various other regulatory factors in global supply chain design had also been studied in Arntzen et al. (1995), Cohen et al. (1989), Goetschalckx et al. (2002), and Oh and Karimi (2004). Supply chain planning without taking taxes into account had been studied in Viswanadham and Balaji (2005a, 2005b). An extended version of this work is also presented in Balaji and Viswanadham (2005).

Even though, the tax integrated model is applicable with more general tax structure, we analyze the model by incorporating tax-holidays enjoyed by locating the various stages of a global supply chain in free trade zones (FTZs). FTZs are special economic zones where export bound goods can be manufactured, assembled and inventoried with generous tax-holidays on custom duty and import/export taxes (Bajpai and Dasgupta (2004)). These zones are introduced in many countries, specifically developing economies, as a part of its export and import (EXIM) policy to encourage exports and FDI on export sector. For the purposes of trade operations, duties and tariffs, the FTZs are considered as a foreign territory. So, any goods supplied to FTZ from Domestic Tariff Area (DTA) are treated as deemed exports and goods brought from FTZ to DTA are treated as imported goods.

Recently, many developing economies in Asia have created FTZs to attract FDI for exports. In Bajpai and Dasgupta (2004) it is observed that China's FDI for the export sector has grown rapidly by the creation of FTZs. It is quite interesting to study the strategic location of the various stages of a global supply chain in the presence of FTZs. In this research we address this strategic problem.

1.2   Organization

Deciding between FDI and outsourcing for various activities of a firm is a hard problem, especially when the number of alternatives to accomplish an activity is many. In Sect. 2, we state this problem. Theoretical models had been developed in the literature to study FDI versus outsourcing (Antras and Helpman (2004), Grossman and Helpman (2002, 2003, 2005), and Helpman et al. (2004)). Even though, these models provide insights in the decision making process, none of them can be applied in the five quantitative context (what percentage to make/source using a particular alternative?). In Sect. 3 we propose a quantitative and weighted Mixed Integer Nonlinear Programming (MINLP) model. This model is a weighted version of the MINLP model for the single product case proposed in Viswanadham and Balaji (2005a). The weighted MINLP model allows the optimal decisions to be obtained by weighing the various objectives. The impact of taxes and tariffs is enormous in the design of a global supply chain. It is critical that the tax consequences and opportunities of introducing business change into the supply chain are included as an integral part of the change process. In Sect. 3 we propose a tax integrated model for optimally deciding the FDI-Outsourcing alternatives for the various stages of a global supply chain. Most supply chain managers already employ various options for sourcing from locations like FTZs to save on customs duties and export/import taxes. In Sect. 4 we analyze the tax integrated model by employing the option of sourcing from FTZs.

2.   Problem Statement and Motivation

A global supply chain spans several countries and regions of the globe. We consider a multi-stage global supply chain network where each stage represents an activity such as, production, assembly, transport, distribution or retail. We assume that the supply chain has N stages, say, $S_1, S_2, \ldots, S_N$. At each stage, the activity could be accomplished using either of the different FDI/Outsourcing alternatives that are possible. For example, in the DEC global supply chain for personal computers of Arntzen et al. (1995), for the demand in UK, the memory manufacturing activity could be accomplished by either of these FDI/Outsourcing alternatives: (a) out-sourcing to a partner in Singapore or Malaysia, or (b) setting up a plant of the company in China to exploit the skilled and low cost labour. Let there be K such different alternatives, $A_1, A_2, \ldots, A_K$ associated with each stage. A 0–1 FDI-Outsourcing strategy, S, is obtained by choosing exactly one FDI/Outsourcing alternative (among the K alternatives) for each stage S_i, $1 \leq i \leq N$. The strategy S can be represented by a $N \times K$ matrix (s_{il}), where $s_{il} = 1$, if for the stage i, alternative l is chosen, $s_{il} = 0$, otherwise. This implies $\sum_{l=1}^{K} s_{il} = 1$, for each stage i. Let the cost matrix (c_{il}) be an $N \times K$ matrix, where c_{il} is the cost associated to the alternative l for the stage i. For a 0–1 FDI-Outsourcing strategy S, the cost $c(S)$ associated with it is defined as, $\sum_{i=1}^{N} \sum_{l=1}^{K} c_{il} \, s_{il}$. An optimal 0–1 FDI-Outsourcing strategy would have the minimum cost. By definition, an optimal 0–1 FDI-Outsourcing strategy minimizes the overall supply chain cost. The problem of deter-mining the optimal 0–1 FDI-Outsourcing strategy is termed as the 0–1 FDI-Outsourcing decision problem.

We consider the relaxed version of the 0–1 strategy, S, in which s_{il} takes the (real) value between 0 and 1 ($0 \leq s_{il} \leq 1$). In this context, the 0–1 FDI-Outsourcing strategy and the 0–1 FDI-Outsourcing decision problem are referred as FDI-Outsourcing strategy and FDI-Outsourcing decision problem, respectively.

For example, we consider the 4-stage supply chain shown in Fig. 1. The system building stage procures PC and software, builds the system, and distribute to the consumers in USA, through a distribution center in USA. The PC procurement stage has two alternatives, namely, procuring from China and Taiwan with procurement costs, 150 USD and 100 USD per unit, respectively. The software procurement stage also has two alternatives, namely, procuring from China and India with procurement costs, 150 USD and 120 USD per unit, respectively. The system building stage could procure using any of these alternatives and build the system in Singapore by outsourcing to a third-party or in Malaysia by establishing a subsidiary. Their respective costs are 100 USD per unit and 110 USD per unit. In this example, we note that the strategy of procuring the PC from Taiwan, software from India and building the system in Singapore, is an optimal 0–1 FDI-Outsourcing strategy.

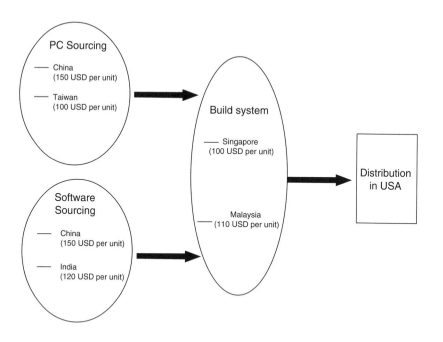

Figure 1. An example to explain 0–1 FDI-outsourcing strategy

As the taxes vary for different tax jurisdictions, by taking into account the tax information for obtaining optimal FDI-Outsourcing strategy would increase the after-tax return of the company. For the example shown in Fig. 1, the optimal 0–1 FDI-outsourcing strategy would be to procure PC from Taiwan, software from India and building the system in Malaysia, if Malaysia has low taxes for building systems. This example gives an insight of how the tax policies and the tax holidays of special economic zones could have an impact on the supply chain strategy. The importance of tax integration in global supply chains can also be realized from the business case discussed in Irving et al. (2005). In the business case considered by Irving et al. (2005), a billion-dollar manufacturing firm has the potential of generating an indexed profit of 271 over the baseline of 100 by locating the key businesses, functions and facilities in tax-favorable locations. However, the firm could realize an indexed profit of 175 over the baseline of 100, in the case in which the business achieves efficiencies and cost savings only by considering logistics, sourcing, manufacturing and services in its global strategy. These motivate us to study the FDI-Outsourcing decision problem by including taxes at the various stages of a supply chain.

3. Modeling

A supply chain could be acyclic or cyclic. The production and distribution networks are examples of acyclic supply chains. The distribution network along with the stage(s) in which the distributed products that are defective are subsequently recalled, repaired, and redistributed, is an example of a cyclic supply chain.

For acyclic supply chains, in this section, we propose MINLP models for the FDI-Outsourcing decision problem. First, we propose a model termed as the weighted base model. We propose an extension of this model by incorporating tax. This model is referred as the tax integrated model. In the proposed models we associate a decision variable, x (without the subscripts taken into consideration), for the production/procurement and inventory activities. This implies that the models have a decision variable x_i associated with the stage i, which could be interpreted as the production (procurement or inventory activity) of a subcomponent i. The decision variable associated to the transport activity between two production stages i and j, is expressed as a function of the decision variables associated to i and j, that is x_i and x_j, and a decision variable y associated to the transport modes. Every stage has production and inventory costs. In the case of FDI the capital costs are absorbed in 8 the production cost. In the case of outsourcing the production cost is equivalent to the procurement cost. The transport cost between the various stages of the supply chain is also captured in the models. The inventory, production and transport costs are assumed to be per lot cost, if their respective lot sizes are specified. Otherwise, the cost corresponds to the per unit cost with lot size set to 1. When the mean demand and the standard deviation of the demand are specified for the final stages (sink nodes[36]) in the supply chain, the mean demand and the standard deviation demand for the non-final stages (non-sink nodes) are computed as follows. Let G be a supply chain network. Let $A(G)$ denote the set of all directed edges (dependencies between the stages) in the supply chain. For a stage i in the supply chain, let μ_i and σ_i be the mean and standard deviation of demand. For a non-sink node i, $\mu_i = \sum_{j:(i,j)\in A(G)} \mu_j$, and, $\sigma_i = \sqrt{\sum_{j:(i,j)\in A(G)} \sigma_j^2}$ σ_i, assuming for all js0 either both μ_j and σ_j are specified (in the case of sink nodes) or computed apriori. This can be achieved by computing μ_i and σ_i for the non-sink nodes in reverse topological order.[37] Assuming that the demand distribution is normal, the demand of stage i is computed as, Di = $\mu_i + k_i$, where k is the service-level.

With these terminologies we propose the weighted base model.

3.1   Weighted Base Model

The weighted base model proposed in this section is a weighted version of the base model proposed in Vishwanadham and Balaji (2005a). For a supply chain network, G, N denotes the number of nodes (stages), and $A(G)$ denotes the set of all directed edges (dependencies between the stages) in the supply chain. The number of possible alternatives at each stage is denoted by K. We propose the following MINLP model termed Weighted Base Model. The objectives production cost (PC), transportation cost (TC), and inventory holding cost (IHC), that have to be minimized are weighted by assigning weights, w_{PC}, w_{TC} and w_{IHC}, respectively. The weights w_t, where $t \in \{PC, TC, IHC\}$, should satisfy, (1) $0 \le w_t \le 1$, and (2) $\sum_t w_t$.

---

[36] A node (or stage) is a sink node if no node depends on it. That is there is no node *j* such that $(i, j) \in A(G)$. A node which is not a sink node is referred as a non-sink node.

[37] A reverse topological ordering is an ordering of the nodes of an acyclic graph such that for any directed arc (*u, v*), *v* appears before *u* in the ordering

MINLP (Weighted Base Model) : minimize $\omega_{PC}(\sum_{i=1}^{N}\sum_{l=1}^{K} PC_{il}\lceil\frac{D_i x_{il}}{PLS_{il}}\rceil)$

$$+\omega_{TC}(\sum_{i=1}^{N}\sum_{l=1}^{K}\sum_{j:(i,j)\in A(G)}\sum_{m=1}^{K}\sum_{r=1}^{n_{mode}} TC_{iljmr}\lceil\frac{D_j y_{iljmr} x_{il} x_{jm}}{TLS_{iljmr}}\rceil)$$

$$+\omega_{IHC}(\sum_{i=1}^{N}\sum_{l=1}^{K} IHC_{il}\lceil\frac{D_i x_{il}}{IHLS_{il}}\rceil(ILT_{il}+PLT_{il}-OLT_{il}))$$

$$\text{subject to} \sum_{l=1}^{K} x_{il} = 1, \forall 1 \le i \le N,$$

$$\sum_{r=1}^{n_{mode}} y_{iljmr} = 1, \forall i,l,j,m, \text{ such that } (i,j) \in A(G),$$

$$OLT_{il}+TT_{iljmr}-ILT_{jm} \le 0, \forall i,l,j,m,r,$$

$$\text{such that } (i,j) \in A(G),$$

$$0 \le x_{il} \le 1, y_{iljmr} = 0 \text{ or } 1, ILT_{jm} \ge 0.$$

In the above model, the decision variables x_{ij}, correspond to the percentage of demand satisfied for a stage i through an alternative l. For any two stages i and j, such that (i,j) 2 $A(G)$, and alternatives l and m, respectively, we define the following for the above model. The terms PC_{il}, TC_{iljmr}, IHC_{il} denote the per lot production cost (PC), transportation cost (TC), and the inventory holding cost (IHC), respectively. The production lot size (PLS), transport lot size (TLS), and inventory holding lot size (IHLS), are denoted by PLS_{il}, TLS_{iljmr} and $IHLS_{il}$, respectively. The number of transport modes available between any two nodes is assumed to be n_{mode}. In a case where a certain transport mode is not available between a pair of nodes, a huge cost could be added with respect to that mode. Since, the weighted base model is a minimization problem this mode would never be included in the optimal solution. It is also assumed that exactly one mode is used to transport goods from stage i to stage j, with alternatives l and m, respectively. This implies, that the decision variables, $y_{iljmr} = 1$, if the goods that has to be transported between stage i and stage j with alternatives l and m, respectively, is transported using the transport mode, r. Otherwise, the decision variables, $y_{iljmr} = 0$. The term, D_i, denotes the demand at stage i. Without loss of generality, D_i, is assumed to be per day demand. For a stage i and an alternative l, the production lead time (PLT), the inbound lead time (ILT) and the outbound lead time (OLT) are denoted by PLT_{il}, ILT_{il}, and

OLT_{il}, respectively. The term TT_{iljmr} denotes the transport time (TT) from i to j with alternatives l and m, respectively, and r is the mode of transport. The terms PLT_{il}, OLT_{il}, ILT_{jm}, and TT_{iljmr}, are assumed to be in days (without loss of generality). The term ILT_{jm} are decision variables in the weighted base model. The decision variables ILT_{jm} should be non-negative, for any non-source node.[38] For source nodes i, ILT_{il} can be set to 0. For a real number α, the term dαe denotes the smallest integer greater than or equal to α.

The objective function of the weighted base model is the weighted sum of the production cost, the inventory cost and the transport cost. The production cost is the sum of the cost of the production lots produced at a stage i using an alternative l. The inventory cost is the sum of 10 the cost of inventory lots inventoried at a stage-alternative combination (i,l). It is computed based on the number of days of inventory that need to be held for (i,l), that is $(ILT_{il} + PLT_{il} - OLT_{il})$. The transport cost is the sum of the cost of the transport lots transported from a stage i to a stage j with their corresponding alternatives l and m, using the transport mode r. For a stage i, the first constraint of the weighted base model should be interpreted as, the sum of the percentage of demand sourced through various alternatives at stage i should sum to 100%. The second constraint is to ensure that exactly one mode of transport is chosen between stage i and j with alternatives l and m, respectively. The third constraint ensures that the inbound lead time of an alternative m at stage j is at least the sum of the outbound lead time of stage i, such that $(i,j) \, 2 \, A(G)$, and the transport time from i to j. This has to hold for all such stages i and its alternatives.

Table 1. Grouping of the input parameters of the weighted base model parameters type

Parameters	Type
D_i	I
$PC_{il}, PLS_{il}, IHC_{il}, IHLS_{il}, PLT_{il}, OLT_{il}$	II
$TC_{iljmr}, TLS_{iljmr}, TT_{iljmr}$	III
w_{PC}, w_{TC}, w_{IHC}	IV

The input parameters to the tax integrated model could be classified as in Table 1, where, (a) Type I denotes the parameter values to be specified for each stage i, (b) Type II denotes the parameter values to be specified for each stage i and its alternative l, (c) Type III denotes the parameter values to

---

[38] A node (or stage) j is a source node, if it is not dependent on any other node. That is there is no node i such that, $(i, j) \, 2 \, A(G)$. A non-source node is a node which is not a source node.

be specified for a combination, stage i with an alternative l, stage j with an alternative m, and transport mode r, and (d) Type IV denotes the parameter values to be specified on production cost, inventory cost and transport cost.

3.2 Tax Integrated Model

Integrating tax in the supply chain decisions may find the company to have a competitive advantage. For example, resourcing or relocating part of the supply chain to a tax advantageous jurisdiction of the globe would certainly make the company to generate huge profits. Therefore, taking the tax information into account can lead to recommend changes in supply chain structure, in sourcing rules, in supplier base, and other factors. It is also better to include tax information at strategic level decision making rather than at a tactical level. In this we propose a strategic decision model by including tax. This model is termed the Tax Integrated Model and obtained by extending the Weighted Base Model.

$$\text{MINLP (Tax Integrated Model)}: \text{minimize } w_{PC}\left(\sum_{i=1}^{N}\sum_{l=1}^{K} PC_{il}\left\lceil\frac{D_i x_{il}}{PLS_{il}}\right\rceil\right)$$

$$+w_{TC}\left(\sum_{i=1}^{N}\sum_{l=1}^{K}\sum_{j:(i,j)\in A(G)}\sum_{m=1}^{K}\sum_{r=1}^{n_{mode}} TC_{iljmr}\left\lceil\frac{D_j y_{iljmr} x_{il} x_{jm}}{TLS_{iljmr}}\right\rceil\right)$$

$$+w_{TAX}\left(\sum_{i=1}^{N}\sum_{l=1}^{K}\sum_{j:(i,j)\in A(G)}\sum_{m=1}^{K}\sum_{r=1}^{n_{mode}} TAX_{iljmr}\left\lceil\frac{D_j y_{iljmr} x_{il} x_{jm}}{TXLS_{iljmr}}\right\rceil\right)$$

$$+w_{IHC}\left(\sum_{i=1}^{N}\sum_{l=1}^{K} IHC_{il}\left\lceil\frac{D_i x_{il}}{IHLS_{il}}\right\rceil(ILT_{il}+PLT_{il}-OLT_{il})\right)$$

$$\text{subject to } \sum_{l=1}^{K} x_{il}=1, \forall 1\leq i\leq N,$$

$$\sum_{r=1}^{n_{mode}} y_{iljmr}=1, \forall i,l,j,m, \text{ such that } (i,j)\in A(G),$$

$$OLT_{il}+TT_{iljmr}-ILT_{jm}\leq 0, \forall i,l,j,m,r,$$

$$\text{such that } (i,j)\in A(G),$$

$$0\leq x_{il}\leq 1, y_{iljmr}=0 \text{ or } 1, ILT_{jm}\geq 0.$$

In the above model TAX_{iljmr} denotes the tax incurred per tax lot (which is denoted by $TXLS_{iljmr}$) for transferring the goods from stage i with alternative l to stage j with alternative m through the transport mode r. The term $^{W_{TAX}}$ is the weight associated with respect to the tax objective. The remaining terms are as defined in the weighted base model. The weights wt, $T \in \{PC, TC, TAX, IHC\}$, are assigned such that (1) $0 \le wt \le 1$, and (2) $\sum_t w_t = 1$.

The objective function of this model is obtained by including tax in the objective function of weighted base model, whereas the constraints are exactly same as in the case of the weighted base model.

The input parameters to the tax integrated model, shown in Table 2, is classified exactly as in the case of weighted base model except to include tax and weight corresponding to it in Type III and Type IV, respectively.

Table 2. Grouping of the input parameters of the tax integrated model parameters type

Parameters	Type
D_i	I
$PC_{il}, PLS_{il}, IHC_{il}, IHLS_{il}, PLT_{il}, OLT_{il}$	II
$TC_{iljmr}, TLS_{iljmr}, TAX_{iljmr}, TXLS_{iljmr}, TT_{iljmr}$	III
$wPC, wTC, wTAX, wIHC$	IV

4. Analysis of the Tax Integrated Model

In this section, we analyze the tax integrated model (TIM) by comparing with the weighted base model (WBM) proposed in Sect. 3, for a 8-stage supply chain shown in Fig. 2. For analysis, we assume a two-country (North and South) model, as in Grossman and Helpman (2003). We also assume the home country of the company to be North. With this assumption, for each stage of the 8-stage supply chain, the different alternatives could be,

1. Outsource South – outsourcing to a low cost country in the South
2. Outsource Home – outsourcing to low cost supplier(s) at home
3. FDI South – FDI in low cost country in the South
4. Home – manufacturing/assembling at home (in-house)

These are referred as alternatives 1–4, respectively. The FDI-Outsourcing decision problem was studied with these alternatives. We grouped the various stages of the 8-stage supply chain as follows.

1. Group 1 – Disk, and Memory manufacturing, Motherboard, and Processor manufacturing

2. Group 3 – Personal Computer assembling
3. Group 4 – Software development
4. Group 5 – System building

We restricted the Groups 1 and 4, to have choices to source only from South. That is Groups 2 and 4 has only two alternatives, Outsource South and FDI South. In the results the non-considered alternatives are denoted by "*X*." Taxes were included in the model with the following assumptions (1) and (2).

1. The activities that are executed in South are assumed to be executed in FTZs. That is for the activities that are accomplished using the alternatives, Outsource South or FDI South, we account for tax-holidays enjoyed by the company by manufacturing/assembling in FTZs. The tax-holidays are taken into account only for North bound demand. No tax-exemption was given to South bound demand as it would be considered an import.

2. The activities that are executed in North are assumed to be executed in Domestic Tariff Areas (DTAs). That is no tax exemption were accounted when the activities were carried out in North. The parameters of the weighted base model and the tax integrated model were set as detailed in the following sub-section.

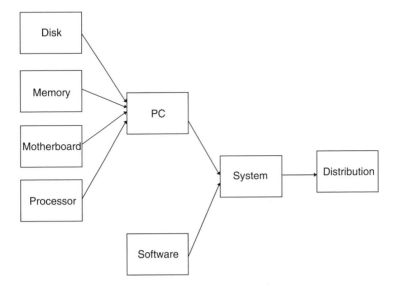

Figure 2. A 8-stage supply chain

4.1   Parameters Setting

Both the models were analyzed for various demand types, namely, High, Medium and Low. For the sink node, Distribution, in the case of High, Medium and Low demand types the mean demand (μ_{Dist}) and standard deviation of demand (σ_{Dist}), are set as follows,

(a) High – μ_{Dist} = 10,000 and σ_{Dist} = 1,000

(b) Medium – μ_{Dist} = 5,000 and σ_{Dist} = 500

(c) Low – μ_{Dist} = 1,000 and σ_{Dist} = 100

By setting the service level to 1, the demand for the various stages with High, Medium, and Low type, are computed as 11,000, 5,500, and 1,100, as detailed in Sect. 3. Production lead time, PLT_{il}, and outbound lead time, OLT_{il}, were set to 1 and 0, respectively, for all i and l. The lot sizes $IHLS_{il}$, PLS_{il}, and TLS_{iljmr}, were set to 1,000,100 and 1,000, respectively. The inventory holding cost associated to the different alternatives with respect to the North and South bound demand, is set for the various stages of the supply chain as follows. The inventory holding cost, IHC_{il}, is set to 1,000 for holding in North, and one-third of its cost, that is 333.33, for holding in South. The production cost, PC_{il}, for the various alternatives, is shown in Table 3. From any stage i to any other stage j, we assumed that there is a single mode of transport, that is n_{mode} = 1. For any two distinct stages, the transport cost, TC_{iljmr}, and the transport time, TT_{iljmr}, from North to South and vice versa, are set to be 1,000 and 2, respectively. Transport cost and transport time within North or South are set to 333.33 (one-third of North–South) and 1 (half of North–South), respectively. The taxes TAX_{iljm} of the tax integrated model were set to 20% of $\left\lceil \dfrac{PC_{il}}{PLS_{il}} \right\rceil$, (1) if l = 2 or 4, or (2) if l = 1 or 3, and m = 1 or 3. Otherwise it is set to 0. The objectives are set equal weights. That is, (a) w_i = 1/3, for all $i \in \{PC, TC, IHC\}$, in the case of weighted base model and (b) w_i = 1/4, for all $i \in \{PC, TC, TAX, IHC\}$, in the case of tax integrated model. With these settings the results obtained by solving the tax integrated model are detailed in the following sub-section.

Table 3. Production cost

Alternative/Demand Type	High	Medium	Low
Outsource South	50	100	150
Outsource Home	150	200	250
FDI South	100	150	200
Home	200	250	300

4.2 Results and Discussion

The models were solved using the CONOPT solver [39] of GAMS Optimization Suite. Both the models were solved for the High, Medium and Low demand cases for North and South bound demand. The optimal FDI-Outsourcing strategies for North – High, Medium and Low demand and South – High, Medium and Low demand, are shown in Tables 4–6 and 7–9, respectively.

In both the models we have the following observations. The results obtained suggest that for both North and South bound demand the optimal strategy is to produce in South. The strategy is quite intuitive as it saves on the production cost and the taxes. We also observe that in both North and South bound demand cases, the percentage of outsourcing decreases and the percentage of FDI increases as we move from the demand type High to Low. This implies that it is cost effective, (1) to outsource when the demand is high, and (2) manufacture inhouse/FDI when the demand is low, as the capital cost would be low. Finally, we observe that the percentage of outsourcing increases and the percentage of FDI decreases as we move from the system building stage to the manufacturing stage of disk, motherboard, memory and processor (in all the cases). This suggests that as we move upstream from the customers the echelons which are closer to the customers should be substantially owned by the company, even though, they may opt to outsource stages that are farther away from the customers.

Table 4. 8-Stage North-High strategy

	Group 1		Group 2		Group 3	
	WBM	TIM	WBM	TIM	WBM	TIM
Outsource South	100	100	100	100	85	85
Outsource Home	X	X	0	0	0	0
FDI South	0	0	0	0	15	15
Home	X	X	0	0	0	0

Table 4. (continued) 8-Stage North-High strategy

	Group 4		Group 5	
	WBM	TIM	WBM	TIM
Outsource South	100	100	50	50
Outsource Home	X	X	0	0
FDI South	0	0	50	50
Home	X	X	0	0

[39] CONOPT is a solver of ARKI Consulting and Development, Denmark, for solving large-scale nonlinear programs (NLPs). More details can be found in http://www.conopt.com

Table 5. 8-Stage North-Medium strategy

	Group 1		Group 2		Group 3	
	WBM	TIM	WBM	TIM	WBM	TIM
Outsource South	100	100	100	100	83	83
Outsource Home	X	X	0	0	0	0
FDI South	0	0	0	0	17	17
Home	X	X	0	0	0	0

Table 5. (continued) 8-Stage North-Medium strategy

	Group 4		Group 5	
	WBM	TIM	WBM	TIM
Outsource South	100	100	50	50
Outsource Home	X	X	0	0
FDI South	0	0	50	50
Home	X	X	0	0

Table 6. 8-Stage North-Low strategy

	Group 1		Group 2		Group 3	
	WBM	TIM	WBM	TIM	WBM	TIM
Outsource South	100	100	100	100	72	72
Outsource Home	X	X	0	0	0	0
FDI South	0	0	0	0	28	28
Home	X	X	0	0	0	0

Table 6. (continued) 8-Stage North-Low strategy

	Group 4		Group 5	
	WBM	TIM	WBM	TIM
Outsource South	100	100	50	50
Outsource Home	X	X	0	0
FDI South	0	0	50	50
Home	X	X	0	0

Table 7. 8-Stage South-High strategy

	Group 1		Group 2		Group 3	
	WBM	TIM	WBM	TIM	WBM	TIM
Outsource South	100	100	100	100	85	92
Outsource Home	X	X	0	0	0	0
FDI South	0	0	0	0	15	8
Home	X	X	0	0	0	0

Table7. (continued) 8-Stage South-High strategy

	Group 4		Group 5	
	WBM	TIM	WBM	TIM
Outsource South	100	100	60	67
Outsource Home	X	X	0	0
FDI South	0	0	40	33
Home	X	X	0	0

Table 8. 8-Stage South-Medium strategy

	Group 1		Group 2		Group 3	
	WBM	TIM	WBM	TIM	WBM	TIM
Outsource South	100	100	100	100	83	89
Outsource Home	X	X	0	0	0	0
FDI South	0	0	0	0	17	11
Home	X	X	0	0	0	0

Table 8. (continued) 8-Stage South-Medium strategy

	Group 4		Group 5	
	WBM	TIM	WBM	TIM
Outsource South	100	100	58	64
Outsource Home	X	X	0	0
FDI South	0	0	42	36
Home	X	X	0	0

Table 9. 8-Stage South-Low strategy

	Group 1		Group 2		Group 3	
	WBM	TIM	WBM	TIM	WBM	TIM
Outsource South	100	100	100	100	72	76
Outsource Home	X	X	0	0	0	0
FDI South	0	0	0	0	28	24
Home	X	X	0	0	0	0

Table 9. (continued) 8-Stage South-Low strategy

	Group 4		Group 5	
	WBM	TIM	WBM	TIM
Outsource South	100	100	50	51
Outsource Home	X	X	0	0
FDI South	0	0	50	49
Home	X	X	0	0

By comparing the weighted base model with the tax integrated model we observe the following. There is no difference in strategy between the two in the case of North bound demand as the tax is 0% when produced in South (FTZ). However, in the case of South bound demand we note that every non-source node (Groups 3 and 5) has a strategic difference as tax of 20% adds to the cost. This subsequently favours outsourcing to FDI as the cost for outsourcing is low.

5. Conclusion

Integrating tax in supply chain decision and locating various supply chain activities in tax advantageous jurisdictions would increase the profitability of a multinational firm. In spite of the importance of this strategic problem, only over the past few years economists and business analysts have started looking at it in an analytical way. However, there were no systematic academic studies on this subject. Our research in this paper is to fill in the gap. In this work we proposed a tax integrated decision model for optimally deciding between Foreign Direct Investment (FDI) and outsourcing at each stage of an acyclic supply chain by taking various tax policies into account. The proposed models were also analyzed empirically for a 8-stage supply chain. Analyzing the models for more realistic data sets and the robustness of the models in those data sets would be an area of research in the future.

References

C. Altomonte, and C. Bonassi. FDI, International Outsourcing and Linkages, CESPRI Working Paper, Centre for Research on Innovation and Internationalisation Processes, Universita' Bocconi, Milano, Italy, 2004 (Downloadable from http://econpapers.repec.org/paper/cricespri/wp155.htm).

B.C. Arntzen, G.G. Brown, T.P. Harrison, and L.L. Trafton. Global supply chain management at digital equipment corporation, *Interfaces* 25 (1), 1995, 69–93.

P. Antras. Firms, contracts, and trade structure, *Quarterly Journal of Economics* 118, 2003, 1375–1418.

P. Antras, and E. Helpman. Global sourcing, *Journal of Political Economy* 112, 2004, 552–580.

N. Bajpai, and N. Dasgupta. Multinational Companies and Foreign Direct Investment in China and India, CSGD Working Paper No.2, January, 2004 (Downloadable from http://www.earthinstitute.columbia.edu/cgsd/ bajpai working papers.html).

K. Balaji, and N. Viswanadham. *A Tax Integrated Approach for Global Supply Chain Network Planning*, submitted to IEEE TASE (2005).

M.A. Cohen, M. Fisher, R. Jaikumar. International manufacturing and distribution networks: a normative model framework, *Managing International Manufacturing*, Ferdows, K., ed., North-Holland, Amsterdam, 1989, p. 67.

S. Domberger. *The Contracting Organization: A Strategic Guide to Outsourcing*, Oxford University Press, Oxford, 1998.

R.C. Feenstra. Integration of trade and disintegration of production in the global economy, *Journal of Economic Perspectives* 12 (Fall), 1998, 31–50.

R.C. Feenstra, and G.H. Hanson. Global production sharing and rising inequality: a survey of trade and wages, NBER Working Paper 8372, 2001 (Downloadable from http://econpapers.repec.org/paper/nbrnberwo/8372.htm).

H. Gorg, A. Hanley, and E. Strobl. Outsourcing, foreign ownership, exporting and productivity: an empirical investigation with plant level data, DIW/GEP Workshop on FDI and International Outsourcing, Berlin, 2004 (Downloadable from http://www.nottingham.ac.uk/economics/staff/details/holger gorg.html).

M. Goetschalckx, C.J. Vidal, and K. Dogan. Modeling and design of global logistics systems: a review of integrated strategic and tactical models and design algorithms, *European Journal of Operational Research* 143 (1), 2002, 1–18.

H.L. F.d. Groot. Macroeconomic consequences of outsourcing: An analysis of growth, welfare, and product variety, *De Economist* 149, 2001, 53–79.

G.M. Grossman, and E. Helpman. Integration versus Outsourcing in industry equilibrium, *The Quarterly Journal of Economics* 117 (1), 2002, 85–120.

G.M. Grossman, and E. Helpman. Outsourcing versus FDI in industry equilibrium, *Journal of European Economic Association* (1), 2003, 317–327.

G.M. Grossman, and E. Helpman. Outsourcing in a global economy, *Review of Economic Studies* 72 (1) (January), 2005, 135–159.

E. Helpman, M. Melitz, and S. Yeaple. Export versus FDI with heterogeneous firms, *American Economic Review* 94, 2004, 300–316.

D. Hummels, J. Ishii, and K. Yi. The nature and growth of vertical specialization in world trade, *Journal of International Economics* 54 (June), 2001, 75–96.

D. Irving, G. Kilponen, R. Markarian, and M. Klitgaard. A tax-aligned approach to SCM, *Supply Chain Management Review* 9 (3) (April), 2005, 57–61.

H.C. Oh, and I.A. Karimi. Regulatory factors and capacity-expansion planning in global chemical supply chains, *Industrial & Engineering Chemistry Research* 43, 2004, 3364–3380.

N. Viswanadham, and K. Balaji. Foreign direct investment or outsourcing: A supply chain decision model, *Proceedings of IEEE-CASE 2005*, Aug 1, 2, 2005, Edmonton, Canada, 2005, pp. 232–237 (2005a).

N. Viswanadham, and K. Balaji. Foreign direct investment or outsourcing : a supply chain decision model, submitted to *European Journal on Operational Research* (2005b).

Integrating Demand and Supply Chains

Puay Guan Goh
GM, SembCorp Logistics, Korea

1. Introduction

The evolution in the implementation of logistics and supply chain management has been rapid in the last few years. As companies move from outsourcing to offshoring of business activities, the ability to coordinate regional or global operations become ever more important, with implications for collaboration, information sharing, planning, and responsiveness in real-time, far-flung supply chains.

At the same time, while many companies have tried to implement some form of collaborative networks, the results from collaboration have been mixed or dubious up until recently. Especially in Asia, collaboration may face even more difficulties related to poor IT and telecommunications infrastructure, lack of trust in information sharing, and insufficient management understanding of collaboration concepts.

As companies continue to push process transformation and system integration inherent in collaborative processes, improvements in shared processes have begun to show. While these efforts may be piecemeal on a project basis, the scope of such smaller-scale projects is clearly more acceptable and more easily implementable in companies than big-bang transformations.

As the integration of processes and information between companies continue to improve, companies improve on their ability to collaborate and work on joint company processes in an extended value chain. With this increasingly close integration, planning and execution functions are no longer separate, but become more coordinated. The ability to integrate the planning and execution functions, close to real-time basis, becomes even more critical.

In addition, it has become increasingly important that both the demand chain and supply chain perspectives are taken into account in this overall process integration. The challenges for companies in balancing the supply and demand chain perspectives could be many. As companies become more far-flung in their global procurement, manufacturing, and sales operations, the ability to coordinate smoothly the entire value chain by matching disparate customer and supplier bases becomes more challenging, and a source of competitor advantage for companies which can execute well. This paper examines different methods for how multinational companies have implemented their inventory management strategies in the Asian context.

2. Challenges in Asia

Within the Asian context, such integration poses particular challenges in the operating environment. The rise of Asia, particularly China in recent years, in low-cost manufacturing has enabled the global footprint of multinationals to spread farther and wider in taking advantage of cost efficiencies. However, the growth and improvement of manufacturing capabilities has not been accompanied by a corresponding improvement in supply chain.

Traditionally, logistics has seen as a backend function, and primarily from the view of asset utilization, with warehousing as a means for storage, and transportation by as trucks operated by small companies (often individual owners) with varying levels of capabilities and number of trucks. Often, this is further hampered by the lack of logistics infrastructure such as roads, ports and airports, skilled workforce, and logistics information technology as well as IT infrastructure nationwide. The relative lack of experience and sophistication in Asian logistics, especially in supply chain, has meant that the emphasis is placed primarily on transactional, perhaps even opportunistic, execution, rather than on planning and coordination over the medium and long-term.

Political and geographical factors also play a part. Asia comprises a number of countries with their own language, cultural, taxation, and historical barriers, and also varying levels of economic development. This makes cross-border coordination more difficult than in the case of large countries such as the United States or economically and legally integrated entities such as the European Union.

Taken together, all these factors have made global coordination difficult to implement seamlessly. Such disparity between Asian operations and the operations in developed countries occurs in ways such as:

- Customers are often based in Europe, concentrating on R&D and US and marketing, while manufacturing sites, or outsourced production, are often based in low cost manufacturing locations in Asia. The ultimate customers of the finished products could be all over the world. This creates challenges in coordination for shipping and logistics, and project management across different geography and time zones.
- Customers are global or regional players with sophisticated management and IT systems while suppliers are small players with low capabilities and resources. Suppliers are less capable of planning optimal inventory and supply chain strategies, or having the resources to support vendor management inventory and collaboration programs.

- Competition is mainly on price due to the pressures of low-cost competition in Asia. As a result, suppliers tend to give lower priority to service levels or quality. At the same time, customers are demanding on service levels (as well as pricing). Service levels for product availability through better inventory, give way to immediate transactional requirements such as price variances between suppliers, creating insufficient focus on joint long-term goals.
- In typical manufacturing environments for end products, the number of input SKUs (Stock Keeping Units) and suppliers are much larger than the number of output SKUs and customers, making inbound logistics into the factory (more probably in Asia) more difficult to coordinate compared to outbound logistics to the end customer.

3.   Supply Chain Strategies

While the challenges are many, the problems are not insurmountable. In order to cope with such disparities between the supply chains and the demand chains, companies must have clear strategies for demand-supply integration and synchronization.

We shall now look at three different strategies that companies in Asia have embarked upon. From there, it may be possible to see different levels of integration, depending on the business needs of the companies, and a logical progression of the extent of integration. The extent of integration would depend on the strategic objectives of the company for supply chain and inventory management, the level of involvement with the supplier base, as well as the level of involvement with the customer base. Especially considering the Asian context that infrastructure and supply chain knowledge and practices could be lacking, it is more important to focus efforts on areas where the payoffs for investments into capabilities, and for transferring knowledge, might be the greatest.

Furthermore, the use of information technology and other logistics-related technologies is often a key cornerstone of such integration initiatives. Only with information sharing, companies can start to build on the information for planning and coordination, and this is critical to the successful implementation of these strategies.

4. Segmentation of Supplier Base for Different Extent of Integration

Not all suppliers are equally important to the operations of the company. The supplier integration effort may be consuming on both time and resources, and analyzing the differences in supplier importance and supplier capabilities enables companies to focus their efforts on the appropriate levels of integration effort for different segments. How might such segmentation strategies be implemented?

In the case of a major semiconductor manufacturer (MSM),[40] segregation of its supplier base is key to effective inventory management. The company is a provider of advanced semiconductor solutions. Its DRAM (Dynamic Random Access Memory) and Flash components are used in advanced computing, networking, and communications products.

The PCBA (Printed Circuit Board Assembly) Operations Division of MSM in Singapore deals with the assembly of RAM chips using TSOP (Thin Small Outline Package), DRAMs, PCBs (Printed Circuit Board) and other discrete components. A typical module consists of a number of memory components that are attached to a printed circuit board. The gold or tin pins on the bottom of the module provide a connection between the module and a socket on another larger PCB. There are also some capacitors and resistors soldered onto the PCB.

Raw materials supplies come from external sources and a single internal source for TSOP. The materials from external sources are broadly classified into two types: generic and customized parts.

Generic parts such as capacitors, resistors, and other discrete components are purchased as consignment goods and there are two main vendors for them. These two suppliers both manufacture and function as trading houses that purchase commodities from other suppliers and manufacturers. Thus, even if they themselves lose production capacity, they can still purchase from other manufacturers to fill in the order, ensuring that supply is stable.

For these goods, auto-replenishment is conducted through supply chain initiatives like negative inventory and VMI (vendor managed inventory). This relieves MSM from the burden of taking care of inventory for consigned goods, leading to large amounts of cost savings. Material availability for these suppliers can be easily seen as they are stored in the same system as MSM. This allows downstream components to view this inventory and plan for output. Monthly conferences are setup between the corporate planning and local planning teams to ensure that suppliers are given the demand forecast.

---

[40] Name has been disguised by request of management.

For specially designed parts, there are five different sources to support global operations, giving a stable supply base for unexpected contingencies. For order processing, the purchase order has to be firstly issued to suppliers. Subsequently, another delivery signal is sent through real-time electronic means, which ensures that deliveries are well-planned and coordinated between supplier and buyer. Its close integration with MSM's internal procurement system helps to facilitate timely information flow and improve the overall performance.

In other words, integration and collaboration is concentrated on the specially designed parts, while responsibility for commodity parts is shifted as far back the supply chain as possible to the suppliers. In that way, inventory costs and administrative costs (and also MSM's control over the process) are reduced for generic products, while ensuring that specialized products receive more attention and process control commensurate to their importance in the production.

5.   Supplier Aggregation

Companies can go a step further in integrating their key suppliers. At some point in time, global sourcing and key supplier initiatives will reach a limit in streamlining the pool of suppliers. In such cases, grouping of components into modules to facilitate modular manufacturing can further help to reduce the complexity of work at the final assembly line. Such modularization of components has been a key aspect of industries that use many small and complex parts, such as the electronics and automotive industry. Ironically, this will create intermediate steps or specialized intermediate players whose job it is to produce these specialized component modules.

Hyundai Mobis, a subsidiary company of Hyundai Motors, provides high-end automotive systems and parts ranging from modules to airbags and driver information systems. Starting in 1999, as a result of the restructuring imperatives from the fallout effects of the Asian financial crisis in 1997 on Korean business conglomerates, Hyundai Mobis began to restructure its business into the core business of automotive parts.

It divested of its rail cars division, heavy machinery business, and auto-mobile division. Hyundai Mobis then focused on pursuing the integration of systems and took on the charge of design, development manufacturing and assembling of all chassis and cockpit modules. In so doing, by focusing on convenience and cost reduction, it has also achieved reduction in weight and number of parts, promptness of assembly, effective inventory management and cost savings.[41]

As a result, Hyundai Mobis has achieved much success in the last 2 years in the automotive parts market, beyond its initial role as a key supplier to Hyundai Motors. In 2004, the company signed a KRW180 billion contract with Daimler Chrysler to provide it with complete chassis modules, while also building a new plant in Alabama for supplying to Daimler Chrysler and other customers in the United States.[42]

When customers order a new Hyundai car, they can choose from different options, such as the cockpit material and the number of air bags. When the choices are made, a list of code numbers for the necessary parts is sent to the Hyundai Mobis module plant for assembly into the appropriate module.[43]

In the era of mass customization, the proliferation of options gives rise to more SKUs, more parts to manage, and more manufacturing line changeovers. While catering to the needs of customers for more options to customize the type of vehicle that they want, Hyundai Motors is able to control its own corresponding increase in the complexity of parts and production management through the effective use of an intermediate supplier. By using modules, it can avoid defects resulting from assembling thousands of small parts. The module provider guarantees the quality of all parts used in the module, reducing the quality assurance and compliance requirements for car assemblers, and thus helping to streamline quality control and manufacturing processes.

In July 2005, it was announced that Hyundai Mobis had embarked on an initiative to track its car components for export using RFID. In a simulated project, car components were tagged with RFID, and tracked from its Ansan factory, export via Busan port, to its final destination in Dubai. This is to enable real-time track and trace of their products, and is expected to be more efficient and cost-effective compared to the use of traditional bar-coding systems.[44]

---

[41] Extracted from Hyundai Mobis corporate website, http://www.mobis.co.kr/eng/

[42] "Mobis Aims to Control Chinese Market", Na, Jeong-ju, Korea Herald, 1st May 2005, http://times.hankooki.com/lpage/biz/200505/kt2005050117300511910.htm

[43] "Hyundai Mobis leads auto-technology market", Kim, So-hyun, Apr 2, 2005, Korea Now, http://koreanow.koreaherald.co.kr/SITE/data/html_dir/2005/04/02/200504020017.asp

[44] "Launch of RFID in Logistics", So, Seong-Hun, Jul 1 2005, ZD Net, translated from Korean, http://www.zdnet.co.kr/news/network/rfid/0,39031109,39137633,00.htm

6.   Active Synchronization of Supply to Demand

The final strategy might involve synchronizing the demand and supply parts of the value chain. This ability to control both the demand and supply parts of the value chain can lead to significant cost savings and value-added capabilities to customers.

Bossard is a global distributor of fasteners and other small component parts (such as screws, bolts, nuts, and rivets), that provides comprehensive engineering and inventory management solutions to equipment manufacturers worldwide. Some of its major customers include ABB, Siemens, John Deere, IBM, Adaptec, Bang & Olufsen, Bosch, Nokia, Celestica, and Flextronics.

In order to serve its customers in their manufacturing expansion in Asia, especially China, Bossard has set up its operations in these countries as well. To meet the customer cost requirements for low cost manufacturing, Bossard had localized its supplier base, many of which are typical small, family-run companies.

Bossard sees its ideal customers as those which see fastener products as either "Uncritical," or "Bottleneck," according to the Kraljic purchasing model.[45] Such customers are more willing to purchase from an aggregator such as Bossard, which simplifies and streamlines the entire process for them. Most of their big customers have short lead time requirements, such as 1 week, while their suppliers may have longer lead time requirements, such as 6 weeks, due to their limited capabilities. The onus has thus fallen on Bossard to hold inventory and assume the risk, in order to serve its customers.[46]

Country business units, with accountability for P&L, mainly do their own planning and procurement individually. As the number of customers, and the number of SKUs handled grew rapidly in its business expansion in the last 5 years, the business units found the lack of visibility with overall demand a potential liability in their inventory management. Due to the lack of visibility between sales, planning and purchasing, purchasing personnel may place orders for large quantities in order to enjoy volume discounts, while sales personnel may.quote volume discount prices while ultimately selling only smaller quantities. As a result, Bossard faces the risk of inventory obsolescence and excessive inventory provisioning in order for its sales people in local business units to cater to the needs of customers.

---

[45] "Purchasing Must Become Supply Management", Kraljic, Peter, Harvard Business Review, September–October 1983, pp. 109–111.

[46] "Supply Chain Management: A Concise Guide", Goh, Puay Guan, Pearson Prentice Hall 2005.

In order to reduce its inventory risk, Bossard has embarked on a planning and collaboration platform. Known as Advanced Planning System Plus (APS+), the APS+ system would act as a cost-based and lead time-based decision support system to determine the impact of purchase decisions on inventory liabilities, costs, and customer fulfillment capabilities. Over a period of 2 years, including planning, system scoping, user feedback, prototyping, testing, and phased rollout through various business units, the system is being implemented and further refined with new user feedback and learning experiences.

Customer sales orders for products are tracked against corresponding purchase orders made to suppliers as well as existing inventory holdings, while optimal inventory holding calculations are made through a combination of sales forecast, lead time, and economic replacement quantities (ERQ). In Bossard's case, ERQ is based on similar calculation to economic order quantity, but with an adjustment for real-world constraints in batch order quantities.

Through real-time visibility, purchase orders are tracked for their potential inventory liability if not delivered to customer, and excess inventory is flagged in management reports for follow-up action. Too many change orders logged by the salespeople, as well as excessive discrepancy between purchase quantity and sales quantity can be highlighted by the system. In addition, any purchases that will lead to excess inventory against forecasted demand and ERQs for customers, will also be highlighted during the placement of the purchase order. In certain cases where inventory rules are violated, management approval is required before the order can proceed. This is a seamless process that automates handovers between departments, and shortens the time and effort for verification of sales and purchase orders prior to approval.

During the implementation period, part numbers are also being standardized so that common parts can be more easily identified and made available for order fulfillment. This would eventually allow for order aggregation and group-based purchasing and order fulfillment for commodity or strategic products.

The APS+ system is shared across its Asia and US operations under its TransPacific division, enabling better global visibility and coordination between sales, customer service, and procurement functions. This ensures that orders are made at the right levels and safety-stock levels are set more accurately, as well as to allow for the transfer of stock holdings allocated to one customer to be used for another customer as the demand situation changes. Through this initiative, it is anticipated that there would be a reduction of Cost of Goods Sold, reduction in inventory obsolescence, and the ability to respond quickly on pricing and availability to customer queries and change orders. It would also lead to reductions in net working capital, with corresponding improvements in balance sheet and value-based financial measures.

7.   Conclusion

It is imperative to remember that supply and demand are dynamic and ever-changing. Therefore, synchronizing the two are always challenging, and require constant vigilant monitoring and adjustment to match supply to demand as far as possible. Within the context of Asia, this is made even more difficult by environmental and business factors.

As such, while collaboration and integration between a company and all its suppliers and customers would be ideal, the reality of the ground situation might give rise to targeted collaborations with a limited number of entities or in environments that are more operationally ready, scaling incrementally with experience. Such targeted collaborations make do with the existing limitations, seek to optimize the situation within the current constraints, and yet can be re-visited when old assumptions and constraints change. At the same time, they also recognize the fundamental reality that not all customers and suppliers are equally important to a company.

In this aspect, companies can adopt a number of different strategies for smoothing over mismatches and variations in demand and supply, especially for their key customers, key product segments, or strategic products. No matter which approach is more preferred or more cost-effective, the ability to create a rapid response to changes in the environment is a critical success factor. It is impossible to expect that a steady-state can exist for long, and companies do well by continually analyzing, adapting, and refining their supply chain strategies based on changes and new information.

Mumbai Tiffin (Dabba) Express

Natarajan Balakrishnan and Chung-Piaw Teo
National University of Singapore, Singapore

1.   Introduction

How would you move 175,000 containers, each with designated origin and destination, on a daily basis, in a reliable and affordable manner? How would you build in features to ensure that the delivery performance remains robust to disruptions in the distribution environment (breakdowns, no-show, new customers being introduced etc.)? Can you do it without using any IT infrastructural investment, and with a workforce that is illiterate? Lest one forgets, the system has to ensure an equivalent reverse flow of empty containers back to the owners every day.

The answers to the above depend on a variety of factors, including social-economical factors and availability of labor forces and conditions of transportation system. In addition to such region specific parameters, the density and relative locations of the origins and destinations, (sparse versus dense layout) also play a major role. For instance, in the postal mail delivery system, the origins and destinations of the mails are normally well spread out within the region. The most common delivery system developed seems to be a **hub and spoke system** – mails will be sent to a central sorting facility for consolidation, which will then be re-directed to the destinations. On the other hand, in most commercial delivery setting, items from a central warehouse need to be delivered to retailers (destinations) spread over the region. A popular way to organize the delivery activities is through the use of a **zoning system** – customers are grouped into zones, each served by a sub-delivery unit. The delivery unit either picks up the items from a designated area within the zone, or direct from the warehouse. The sorting is normally done near the origins (central warehouse) (Fig. 1).

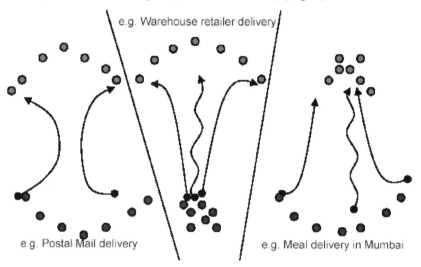

Figure 1. Sparsity/density of origins/destinations affects the choice of delivery organization

In this chapter, we take a close look at a third kind of delivery problem, where the origin is spread over a region, with 70% of the destinations clustering near a central area. This is a feature in a meal delivery system in the sprawling metropolis of Mumbai, where although many workers commute into the city to work, they still want to enjoy home-cooked meals for lunch. There is currently a highly e.g. Warehouse retailer delivery e.g. Postal Mail delivery e.g. Meal delivery in Mumbai efficient and low cost delivery solution, the Dabbawallah System, (roughly translated as "dabba" – the container for carrying food and "wallah" – the people, in this case the people who transport the containers) which has astounded the logistics professionals all over the world (Forbes, 2001). Such is the popularity of a system that it is as famous a tourist attraction as the Gateway of India in Mumbai. The system, which has been in operation for over 100 years, has been so impressive in its working, that a visit to see the Mumbai Tiffin Express in operation was specially requested for when Prince Charles visited Mumbai in 2003!

How do the dabbawallahs, a group of lowly educated Indians from the Pune region, build a world class distribution system to solve this delivery problem?

2. The System: How it Operates

The system operates around meticulous timing and coordinated team-work. Figure 2 shows the time chart of a typical daily operation.

The system operates using a zoning system approach. Each zone is served by a team of 20–25 dabbawallahs, each serving around 30 customers per day. Each team operates as a separate business unit, and the team leader (called mukadam) is responsible for the efficient coordinated functioning of the team. The teams are thus self-administered work units sharing common agenda with each other. The Dabbawallah System epitomizes a self-regulating decentralized delivery system, loosely organized as a cooperative system, under the *Nutan Tiffin Box Suppliers Association.* There are only three hierarchies of authority, making it a rather flat organization. This includes some 5,000 workers, 800 mukadams, and a small number of Executive Committee in the Association. The Executive Committ-ee is primarily involved in conflict resolution, setting the agenda and administering the welfare activities.

Each customer is charged around 175 rupees (around US$4) per month for the service. The customer only needs to invest upfront a token sum to purchase the dabba to store the meals. To subscribe to the service, one merely has to approach any dabbawallah and provide the home address

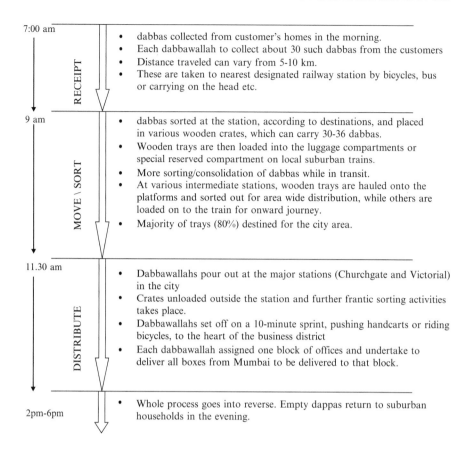

Figure 2. The time chart of the dabbawallah operations

where the service would be required. This information will be immediately disseminated to the team responsible for the delivery in the zone the customer resided in. Further negotiation of price and delivery timings will be done between the customer and the team leader.

After paying $1.25 per crate and $2.50 per man per month to the Western Railway for transport, the money collected from the customers by the dabbawallah goes into the cooperative pool that he belongs to. Out of the accumulated fund, he is paid a monthly salary of $70. Of that sum collected from customers, $0.25 goes to the parent association. The association, after minimal expenses, spends the rest of the $1,000 every month to a charitable trust.

Figure 3 shows some estimates of dabbas flowing through several selected stations along the railway. The data was collected in early 2003.

3.   The System: Secrets of Success

Much of the success of the Dabbawallah System can be attributed to the **committed workforce**. Ninety percent of the city's approximately 5,000 dabbawallahs come principally from the Pune region of the state of Maharashtra, several hours from Mumbai. All have left poor farming communities in search of a means to support their families. They share a common language and have **strong social bond**. They take great pride in what they do, and understand that their livelihood depends on their ability to deliver the meals efficiently, come what may. On the office wall of the Tiffin Carriers association hangs a list of 23 rules, a corporate code of conduct. One in particular sums up the dabbawallah's ethic: "No customer should go without food."

The delivery system builds around the **extensive commuter railway** system, the backbone of the Mumbai transport network, connecting the vast suburbs to the city areas. The timetable of the railway system and the common delivery deadline for the tiffins induce a natural **clockspeed** into the delivery operations. Every dabbawallah understands the need to race against time to reach his destined station to meet his badli (counterpart) who will board the local train with his quota at the precise hour at a given station. The effect of slip delivery is immediately discernable.

Much of the credits for the success must also go to the fact that the system seeks to evolve continuously to **adapt to local conditions**, rather than blindly following best-practices imported from elsewhere. This is evident, for instance, from the evolution of the **coding system** used in the system to track the flow of dabbas within the entire delivery network. While bar codes (RFID in the future?) are common to modern day delivery system, its high cost (relative to cost of the service) and the environment (manned by illiterate workers) meant that the system has to adapt a new approach to track the flow of dabbas. How do they do it? The dabbawallahs chose to evolve a coding system that "speaks" to its bunch of illiterate workers, fully recognizing the fact that its strength lies on its cheap labour and committed workforce. The code, which is painted on the dabba top, is restricted first by the size of the top itself – 6 in. in diameter. The code uses colour, dashes, crosses, dots and simple symbols to indicate the various parameters like originating suburb, route to take, destination-station, whose responsibility, the street, building, floor et al. The system by its simple structure ensures a smooth flow to and from destination, though a dabba might pass through as many as six persons in each direction of movement everyday. Since the system is operated by strictly controlled but loosely linked groups, each group has a certain amount of flexibility in personalizing the coding system.

Thus the mukadam, the manager of each operating group, can personalize some colors etc to differentiate the dabbas pertaining to each of his group dabbawallah. Over the years, dashes and crosses have been replaced by simple text which is easy to read (See Fig. 4).

4.   The New Environment

The business of the dabbawallahs had gone through several major upheavals in recent decades. A substantial chunk of business was lost with the closure of the textile mills in Mumbai, and when subsidized in-house canteens started to appear in schools and food coupon system in corporates. With changing times, Mumbai's young executives have also developed affluent life-styles with the wife holding another full time job. Further changes in the operating environment came about when business activities began to move to the new Mumbai business district inland over the last few years and thus complicating the already intricate distribution network.

The dabbawallahs have responded to each of these threats by constantly adapting their business practices. They have continuously innovated, offering new services, while riding on their core strength of on-time and reliable delivery services. To quote a few – a new express delivery service (pick-up at 11 a.m.) has been introduced in recent years; a linkage with groups of housewives to cook and supply dabbas for customers who prefer home-cooked food; they have even worked with marketing agents like the brand management team at Lowe Lintas, Mumbai, to distribute free Surf Excel-branded tablemats, along with the dabbas, to the offices.

A sign that the system is evolving is clearly displayed in Fig. 5. The standard dabbas used for the meal delivery have slowly given way to more elaborate and colorful packaging, and the system has cleverly evolved to accommodate the request of customers who want their meals to be delivered to them in a fancy and "appetizing" manner.

How long can this devotion to home cooked food remain among customers, no one can tell. But we can sure count on the dabbawallahs to continue to strive and evolve with the new economic conditions.

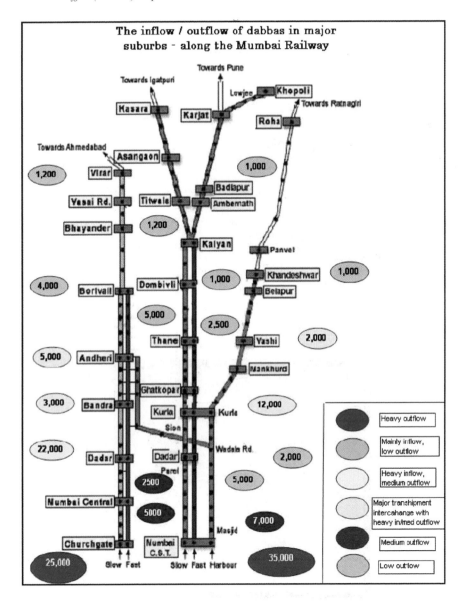

Figure 3. The inflow/outflow of dabbas in major suburbs along the Mumbai railway

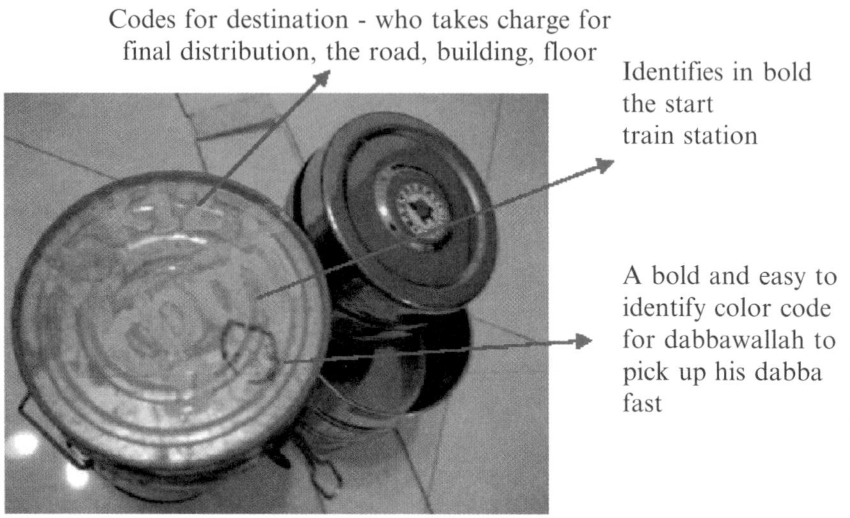

Codes for destination - who takes charge for
final distribution, the road, building, floor

Identifies in bold
the start
train station

A bold and easy to
identify color code
for dabbawallah to
pick up his dabba
fast

Figure 4. A common coding system in use

Figure 5. Dabbas with new designs

Index

A
Annual volume, 36, 41, 42
Apparel retailing industry, 106, 107

B
Bass diffusion model, 180, 195–196
Berth allocation planning, 70, 72–79, 87,
    90, 93, 95–97, 102
Berthing decision, 79, 89
Berth-Time-Requested (BTR), 70

C
Cause consequence diagram, 212–213,
    215, 217
Chemical supply chains, 47–50, 58–62, 67
Competitive products, 112
Container terminal, 70, 76, 78
Consumer market, 2, 18, 149, 155, 158,
    162
Coordination, 25, 30, 225, 230, 263, 264,
    269
Complementary products, 112–113
Cosmetic distribution channels, 5, 14–20,
    25
Cosmetic industry of China, 1–25
Cosmetic products, 7–12, 14, 16
Cosmetic supply chain, 4, 5, 9–14, 24, 25
Cosmetic supply chain configurations,
    9–14, 23, 25

D
Delay cost, 58, 79
Demand forecasting, 9, 49–50, 63, 107,
    109, 113, 125, 127, 133–135, 137, 143,
    163–165, 224–226, 231, 265, 269
Deviation, 73, 89, 135, 136, 186,
    206–211, 214, 215, 218, 221, 248, 254
Disaster, 204–208
Discrete maximum pressure, 74, 90–93,
    96, 100–102
Disruption, 21, 66, 70, 204, 206–209,
    212–214, 217, 218, 220, 221, 272
Distribution configurations, 20–22, 25
Distribution cost, 16–18, 33–35
Divestment, 52, 53, 55
Dynamic supply network, 23–24

E
E-commerce, 148
Enterprise Resource Planning, 55, 203
Environmental awareness, 57–58
Estimated-Time-of-Arrival (ETA), 70, 77
Export, 10, 12–14, 19, 170, 172, 174,
    180, 189, 191–193, 202, 221, 243–245,
    267

F
Foreign Direct Investment, 172, 241–258

G
Global chemical industry, 46, 53, 57, 60,
    66
Global diffusion, 170–175, 180, 194, 195
Global supply chains, 65, 66, 170,
    173–175, 193, 194, 202, 204, 221,
    243–245, 247
Government regulations, 16

I
International Council of Chemical
    Associations, 57
Inventory management, 62, 109, 124,
    126, 205, 262, 264, 265, 267, 268
Inventory replenishment, 124–142
IP protection, 10, 11
ISO 9000, 170–177, 179–185, 187, 188,
    190, 193–195, 197
IT systems, 54–56, 263

K
k-median clustering, 120, 122, 123, 143

L
Limited life-cycle retail products, 124
Location cost, 78

M
Make-to-stock, 49
Manufacturing operations in chemical,
    47–50, 55
Marketing strategy, 3

Market readiness, 4
Merchandise planning, 106, 109, 118, 144
Merchandise testing, 109–123, 143
Mergers & Acquisitiions, 25, 35, 52, 53, 55, 60
Multinational corporations, 2
Mumbai dabbawallahs, 273–278

N
Neighborhood search, 84–87
Newsvendor problem, 124, 149

O
Offshore manufacturing, 10, 12–14
Online auctions, 51, 162
Optimization, 42, 56, 63, 64, 73, 75, 76, 79, 93, 109, 112, 114, 116, 117, 124–126, 130, 141, 142, 148, 210, 243, 255
Outsourced manufacturing, 2, 10, 12–14, 206, 263
Outsourcing, 12, 57, 203, 242–247, 252, 255, 258, 262

P
Power, 66, 212, 225–229, 231–235
Product diversion, 153–157
Product rollovers, 24
Production planning of chemical plants, 49

R
Regional analysis, 5–6
Relative savings, 40–42
Returns of merchandise, 132–133
Risk management, 203–205, 208–211, 213–215
Risks, 21, 61–62, 66, 148, 203–206, 221, 243
Rolling horizon birth allocation, 71, 74, 78, 87–89, 92, 95–101

S
Secondary market, 148–150, 152–165, 167
Sequence pair concept, 73, 76, 79–81
Sourcing, 47–48, 56, 60, 62, 63, 108, 204, 208, 209, 243, 245, 247, 251, 266
Supplier aggregation, 266–267
Supply chain collaboration, 30–33, 36, 37, 40, 42, 43, 224–226, 230, 232–235
Supply chain exceptions, 205, 211–212
Supply chain integration, 224, 226, 244, 247, 262, 264
Supply chain risks, 202–206, 213
Supply chain strategies, 3, 10, 204, 205, 234, 247, 263, 264, 270

T
Tax, 7, 64, 120, 122, 139, 242–245, 247, 250–255, 258
Test store selection, 115, 118, 119, 122
Transportation, 11–13, 23, 25, 31, 33–35, 38, 39, 41–43, 47, 50, 56, 60, 63, 64, 165, 204, 206, 207, 248, 249, 263, 272
Trust, 204, 212, 225–227, 229–235, 262, 274

U
U. S. Grocery Industry, 31, 32

V
Virtual wharf mark, 73, 83, 84, 87, 94, 102

W
Women's specialty apparel retailer, 106, 118–123

Early Titles in the
INTERNATIONAL SERIES IN
OPERATIONS RESEARCH & MANAGEMENT SCIENCE
Frederick S. Hillier, Series Editor, *Stanford University*

Saigal/ *A MODERN APPROACH TO LINEAR PROGRAMMING*
Nagurney/ *PROJECTED DYNAMICAL SYSTEMS & VARIATIONAL INEQUALITIES WITH APPLICATIONS*
Padberg & Rijal/ *LOCATION, SCHEDULING, DESIGN AND INTEGER PROGRAMMING*
Vanderbei/ *LINEAR PROGRAMMING*
Jaiswal/ *MILITARY OPERATIONS RESEARCH*
Gal & Greenberg/ *ADVANCES IN SENSITIVITY ANALYSIS & PARAMETRIC PROGRAMMING*
Prabhu/ *FOUNDATIONS OF QUEUEING THEORY*
Fang, Rajasekera & Tsao/ *ENTROPY OPTIMIZATION & MATHEMATICAL PROGRAMMING*
Yu/ *OR IN THE AIRLINE INDUSTRY*
Ho & Tang/ *PRODUCT VARIETY MANAGEMENT*
El-Taha & Stidham/ *SAMPLE-PATH ANALYSIS OF QUEUEING SYSTEMS*
Miettinen/ *NONLINEAR MULTIOBJECTIVE OPTIMIZATION*
Chao & Huntington/ *DESIGNING COMPETITIVE ELECTRICITY MARKETS*
Weglarz/ *PROJECT SCHEDULING: RECENT TRENDS & RESULTS*
Sahin & Polatoglu/ *QUALITY, WARRANTY AND PREVENTIVE MAINTENANCE*
Tavares/ *ADVANCES MODELS FOR PROJECT MANAGEMENT*
Tayur, Ganeshan & Magazine/ *QUANTITATIVE MODELS FOR SUPPLY CHAIN MANAGEMENT*
Weyant, J./ *ENERGY AND ENVIRONMENTAL POLICY MODELING*
Shanthikumar, J.G. & Sumita, U./ *APPLIED PROBABILITY AND STOCHASTIC PROCESSES*
Liu, B. & Esogbue, A.O./ *DECISION CRITERIA AND OPTIMAL INVENTORY PROCESSES*
Gal, T., Stewart, T.J., Hanne, T. / *MULTICRITERIA DECISION MAKING: Advances in MCDM Models, Algorithms, Theory, and Applications*
Fox, B.L. / *STRATEGIES FOR QUASI-MONTE CARLO*
Hall, R.W. / *HANDBOOK OF TRANSPORTATION SCIENCE*
Grassman, W.K. / *COMPUTATIONAL PROBABILITY*
Pomerol, J-C. & Barba-Romero, S. / *MULTICRITERION DECISION IN MANAGEMENT*
Axsäter, S. / *INVENTORY CONTROL*
Wolkowicz, H., Saigal, R., & Vandenberghe, L. / *HANDBOOK OF SEMI-DEFINITE PROGRAMMING: Theory, Algorithms, and Applications*
Hobbs, B.F. & Meier, P. / *ENERGY DECISIONS AND THE ENVIRONMENT: A Guide to the Use of Multicriteria Methods*
Dar-El, E. / *HUMAN LEARNING: From Learning Curves to Learning Organizations*
Armstrong, J.S. / *PRINCIPLES OF FORECASTING: A Handbook for Researchers and Practitioners*
Balsamo, S., Personé, V., & Onvural, R./ *ANALYSIS OF QUEUEING NETWORKS WITH BLOCKING*
Bouyssou, D. et al. / *EVALUATION AND DECISION MODELS: A Critical Perspective*
Hanne, T. / *INTELLIGENT STRATEGIES FOR META MULTIPLE CRITERIA DECISION MAKING*
Saaty, T. & Vargas, L. / *MODELS, METHODS, CONCEPTS and APPLICATIONS OF THE ANALYTIC HIERARCHY PROCESS*
Chatterjee, K. & Samuelson, W. / *GAME THEORY AND BUSINESS APPLICATIONS*
Hobbs, B. et al. / *THE NEXT GENERATION OF ELECTRIC POWER UNIT COMMITMENT MODELS*
Vanderbei, R.J. / *LINEAR PROGRAMMING: Foundations and Extensions, 2nd Ed.*
Kimms, A. / *MATHEMATICAL PROGRAMMING AND FINANCIAL OBJECTIVES FOR SCHEDULING PROJECTS*
Baptiste, P., Le Pape, C. & Nuijten, W. / *CONSTRAINT-BASED SCHEDULING*
Feinberg, E. & Shwartz, A. / *HANDBOOK OF MARKOV DECISION PROCESSES: Methods and Applications*
Ramík, J. & Vlach, M. / *GENERALIZED CONCAVITY IN FUZZY OPTIMIZATION AND DECISION ANALYSIS*
Song, J. & Yao, D. / *SUPPLY CHAIN STRUCTURES: Coordination, Information and Optimization*
Kozan, E. & Ohuchi, A. / *OPERATIONS RESEARCH/ MANAGEMENT SCIENCE AT WORK*

Early Titles in the
INTERNATIONAL SERIES IN
OPERATIONS RESEARCH & MANAGEMENT SCIENCE
(Continued)

Bouyssou et al. / *AIDING DECISIONS WITH MULTIPLE CRITERIA: Essays in
    Honor of Bernard Roy*
Cox, Louis Anthony, Jr. / *RISK ANALYSIS: Foundations, Models and Methods*
Dror, M., L'Ecuyer, P. & Szidarovszky, F. / *MODELING UNCERTAINTY: An Examination
    of Stochastic Theory, Methods, and Applications*
Dokuchaev, N. / *DYNAMIC PORTFOLIO STRATEGIES: Quantitative Methods and Empirical Rules
    for Incomplete Information*
Sarker, R., Mohammadian, M. & Yao, X. / *EVOLUTIONARY OPTIMIZATION*
Demeulemeester, R. & Herroelen, W. / *PROJECT SCHEDULING: A Research Handbook*
Gazis, D.C. / *TRAFFIC THEORY*
Zhu/ *QUANTITATIVE MODELS FOR PERFORMANCE EVALUATION AND BENCHMARKING*
Ehrgott & Gandibleux/ *MULTIPLE CRITERIA OPTIMIZATION: State of the Art Annotated Bibliographical
    Surveys*
Bienstock/ *Potential Function Methods for Approx. Solving Linear Programming Problems*
Matsatsinis & Siskos/ *INTELLIGENT SUPPORT SYSTEMS FOR MARKETING
    DECISIONS*
Alpern & Gal/ *THE THEORY OF SEARCH GAMES AND RENDEZVOUS*
Hall/*HANDBOOK OF TRANSPORTATION SCIENCE - 2nd Ed.*
Glover & Kochenberger/ *HANDBOOK OF METAHEURISTICS*
Graves & Ringuest/ *MODELS AND METHODS FOR PROJECT SELECTION:
    Concepts from Management Science, Finance and Information Technology*
Hassin & Haviv/ *TO QUEUE OR NOT TO QUEUE: Equilibrium Behavior in Queueing Systems*
Gershwin et al/ *ANALYSIS & MODELING OF MANUFACTURING SYSTEMS*
Maros/ *COMPUTATIONAL TECHNIQUES OF THE SIMPLEX METHOD*
Harrison, Lee & Neale/ *THE PRACTICE OF SUPPLY CHAIN MANAGEMENT: Where Theory and
    Application Converge*
Shanthikumar, Yao & Zijm/ *STOCHASTIC MODELING AND OPTIMIZATION OF
    MANUFACTURING SYSTEMS AND SUPPLY CHAINS*
Nabrzyski, Schopf & Węglarz/ *GRID RESOURCE MANAGEMENT: State of the Art and Future Trends*
Thissen & Herder/ *CRITICAL INFRASTRUCTURES: State of the Art in Research and Application*
Carlsson, Fedrizzi, & Fullér/ *FUZZY LOGIC IN MANAGEMENT*
Soyer, Mazzuchi & Singpurwalla/ *MATHEMATICAL RELIABILITY: An Expository Perspective*
Chakravarty & Eliashberg/ *MANAGING BUSINESS INTERFACES: Marketing, Engineering, and
    Manufacturing Perspectives*
Talluri & van Ryzin/ *THE THEORY AND PRACTICE OF REVENUE MANAGEMENT*
Kavadias & Loch/*PROJECT SELECTION UNDER UNCERTAINTY: Dynamically Allocating Resources to
    Maximize Value*
Brandeau, Sainfort & Pierskalla/ *OPERATIONS RESEARCH AND HEALTH CARE: A Handbook of
    Methods and Applications*
Cooper, Seiford & Zhu/ *HANDBOOK OF DATA ENVELOPMENT ANALYSIS: Models and Methods*
Luenberger/ *LINEAR AND NONLINEAR PROGRAMMING, 2nd Ed.*
Sherbrooke/ *OPTIMAL INVENTORY MODELING OF SYSTEMS: Multi-Echelon Techniques,
    Second Edition*
Chu, Leung, Hui & Cheung/ *4th PARTY CYBER LOGISTICS FOR AIR CARGO*
Simchi-Levi, Wu & Shen/ *HANDBOOK OF QUANTITATIVE SUPPLY CHAIN ANALYSIS: Modeling
    in the E-Business Era*
Gass & Assad/ *AN ANNOTATED TIMELINE OF OPERATIONS RESEARCH: An Informal History*
Greenberg/ *TUTORIALS ON EMERGING METHODOLOGIES AND APPLICATIONS IN OPERATIONS
    RESEARCH*
Weber/ *UNCERTAINTY IN THE ELECTRIC POWER INDUSTRY: Methods and Models for Decision
    Support*
Figueira, Greco & Ehrgott/ *MULTIPLE CRITERIA DECISION ANALYSIS: State of the Art Surveys*

Early Titles in the
INTERNATIONAL SERIES IN
OPERATIONS RESEARCH & MANAGEMENT SCIENCE
(Continued)

Reveliotis/ *REAL-TIME MANAGEMENT OF RESOURCE ALLOCATIONS SYSTEMS: A Discrete Event Systems Approach*
Kall & Mayer/ *STOCHASTIC LINEAR PROGRAMMING: Models, Theory, and Computation*
Sethi, Yan & Zhang/ *INVENTORY AND SUPPLY CHAIN MANAGEMENT WITH FORECAST UPDATES*
Cox/ *QUANTITATIVE HEALTH RISK ANALYSIS METHODS: Modeling the Human Health Impacts of Antibiotics Used in Food Animals*
Ching & Ng/ *MARKOV CHAINS: Models, Algorithms and Applications*
Li & Sun/ *NONLINEAR INTEGER PROGRAMMING*
Kaliszewski/ *SOFT COMPUTING FOR COMPLEX MULTIPLE CRITERIA DECISION MAKING*

 * A list of the more recent publications in the series is at the front of the book *

Printed in the United States of America